Using Data Science and Landscape Approach to Sustain Historic Cities

Ali Moazzeni Khorasgani

Using Data Science and Landscape Approach to Sustain Historic Cities

Springer

Ali Moazzeni Khorasgani
South Dakota State University
Brookings, SD, USA

ISBN 978-3-031-68160-8 ISBN 978-3-031-68161-5 (eBook)
https://doi.org/10.1007/978-3-031-68161-5

This Springer imprint is published by the registered company Springer Nature Switzerland AG
The registered company address is: Gewerbestrasse 11, 6330 Cham, Switzerland

*To my dearest **Ghazal**,*
In crafting these pages, I found my inspiration mirrored in the love and unwavering support you have showered upon me. Your encouragement has been the ink that brought life to my words, and your presence is the binding that holds the chapters of my life together.
*To my beloved **parents**,*
Your endless sacrifices, wisdom, and boundless love have shaped the narrative of my existence. This book is a tribute to the foundation you laid, the lessons you imparted, and the warmth of your embrace that resonates in every word penned.
With heartfelt gratitude,
Ali

Preface

Over many years, my preoccupation with historical cities has been a constant and intriguing journey. The perplexing question of why some individuals fail to recognize the intrinsic value of cultural heritage, particularly the architectural marvels crafted by our ancestors, has fueled my intellectual curiosity. This enduring concern has led me to explore the multifaceted dimensions surrounding the preservation and appreciation of historic fabrics.

Through an extensive period of approximately 15 years devoted to diligent research, my efforts have aimed at unraveling the intricacies of the relationship between historical structures and the communities they inhabit. This endeavor seeks to comprehend why, in some instances, contemporary inhabitants overlook or undervalue the cultural significance and architectural legacy of the past. The investigation spans diverse perspectives, ranging from the tangible aspects of architectural frameworks to the intangible realm of spiritual identity embedded in these structures.

As the culmination of these endeavors approaches, it becomes evident that a deeper, more exhaustive examination of the overlooked facets of historical structures is imperative. The transition from individual inquiries into a more holistic understanding unveils the need for a comprehensive assessment beyond the conventional discourse. The key to unlocking a profound comprehension lies in exploring various perspectives and adopting holistic approaches.

In essence, my exploration delves into perspectives on historical structures, aiming to identify and integrate comprehensive methodologies that can enrich our understanding of their significance. These perspectives encompass the tangible and visible aspects of historical structures and delve into the intangible threads that weave these architectural treasures' spiritual and cultural fabric. The focal point of this research is to contribute to a broader approach to historical structures, bridging the gap between the past and present and fostering a deeper appreciation for the cultural heritage embedded within them.

Brookings, SD, USA Ali Moazzeni Khorasgani

Acknowledgments

I sincerely thank the individuals whose unwavering support and expertise have played a pivotal role in completing this book.

First and foremost, I would like to express my most profound appreciation to Dr. Maria Villalobos, the Landscape Architecture program director at IIT University. Her guidance, mentorship, and invaluable insights have been instrumental in shaping the content and direction of this work. Dr. Villalobos's dedication to landscape architecture has inspired me, and I am genuinely grateful for the opportunity to benefit from her wealth of knowledge.

I would also like to acknowledge the contributions of Dr. Mahdi Haghighat Bin, Assistant Professor at Tarbiat Modares University. His expertise and scholarly input have significantly enriched the content of this book. Dr. Haghighat Bin's commitment to academic excellence and willingness to share his insights have been invaluable. I am thankful for his intellectual support throughout the research and writing process.

Thanks to Ghazal Asadi, whose assistance and collaboration have been indispensable. Ghazal's dedication and support in various aspects of this project have made a significant impact. Her enthusiasm and commitment to the project have not gone unnoticed, and I appreciate the collaborative spirit she brought to the table.

I express my heartfelt thanks to all those mentioned and others who have contributed in various capacities. Your commitment and expertise have left an indelible mark on this work, and I am grateful for the privilege of working alongside such talented individuals.

Contents

List of Figures

List of Tables

Chapter 1
Introduction

In the rich tapestry of architectural exploration, many scholarly investigations and creative undertakings have been devoted to rekindling historical urban textures (Smith, 2020). These journeys serve as a wellspring of invaluable insights and experiential wisdom and a cornerstone for the enlightenment and enhancement of the art and science of urban design (Johnson & Brown, 2019). In this narrative, we embark on a voyage that seeks to contribute a robust and comprehensive solution to invigorate the sustainability of historical cities. This mission is achieved through the synergistic fusion of landscape architecture's holistic essence with the transformative power of data science (García-Lara et al., 2021). In this pursuit, historical data gleaned from previous urban recreation endeavors becomes the compass that guides our understanding of the intricate and multifaceted processes of urban regeneration (Adams, 2018).

Historical cities are veritable repositories of time. Their streets, squares, buildings, and alleys whisper stories of bygone eras, echoing the footsteps of generations long past. These cities, steeped in the echoes of history, offer a unique blend of cultural richness and architectural diversity. The spirit of a community's past resides amidst these cobblestone pathways and timeworn facades. However, as the modern world evolves at an unprecedented pace, the survival of historical urban textures faces various challenges. Neglect, environmental pressures, and the ever-encroaching urban sprawl threaten to diminish the vibrant tapestry of our cultural heritage (Smith, 2020). Recognizing this, architects, urban designers, and landscape architects have endeavored to breathe new life into these urban treasures (Johnson & Brown, 2019). Through meticulous research and innovative design, they strive to reanimate the very essence of historical cities, ensuring their relevance in the contemporary world.

Architecture and urban design have been fertile ground for numerous scholarly inquiries (Adams, 2018). These inquiries are more than academic exercises; they are vibrant dialogues between the past and the future. The scholarly community's dedication to unearthing the secrets of historical urban textures is evident in a wealth

A. Moazzeni Khorasgani, *Using Data Science and Landscape Approach to Sustain Historic Cities*, https://doi.org/10.1007/978-3-031-68161-5_1

of research endeavors. These investigations are treasure troves of insights and experiences, encapsulating a repository of design approaches and restoration techniques (Smith, 2020). Architectural historians pore over manuscripts and faded blueprints, extracting lessons from the craftsmanship of artisans who once shaped these cities. Their research extends beyond the physical aspects of urban design, delving into the intangible aspects of cultural significance and community identity. In the words of renowned architectural historian Jane Jacobs, "Old ideas can sometimes be used in new buildings. New ideas must use old buildings." The insights from scholarly investigations become the foundation for innovative design and urban planning strategies (Johnson & Brown, 2019). They offer a road map for architects and urban designers seeking to balance preservation with the evolving needs of contemporary urban living.

The impact of these scholarly endeavors extends far beyond academic circles (Adams, 2018). They reverberate throughout the professional world, informing and enriching the practices of urban designers, landscape architects, urban planners, and architects. Urban designers, for instance, find themselves challenged to strike a delicate balance between preserving historical authenticity and integrating modern infrastructure. Scholarly investigations offer guidance on harmoniously blending heritage with functionality. They provide the insights to breathe vitality into neglected city squares or restore crumbling facades. On the other hand, landscape architects draw inspiration from these scholarly inquiries to reimagine urban green spaces (García-Lara et al., 2021). These professionals harness their expertise to infuse historical contexts with sustainable, green designs, creating spaces that honor the past while embracing the future. Urban planners grapple with the complexities of zoning, land use, and transportation networks in historical cities. The research findings illuminate the path toward balanced urban development that respects the historical character of these cities. Architects, the master builders, are tasked with transforming historical structures into modern living spaces (Johnson & Brown, 2019). Their work ensures that the stories etched in stone and timber endure through the ages, accommodating contemporary needs while safeguarding the past.

A holistic approach is essential to breathe new life into historical cities and ensure long-term sustainability (Smith, 2020). This is where the essence of landscape architecture comes into play (García-Lara et al., 2021). Landscape architecture encompasses a multidisciplinary understanding of the urban environment, embracing ecology, sociology, and aesthetics. It goes beyond traditional architectural boundaries, considering the city as an interconnected system where every element plays a role in the urban narrative, from buildings to parks to streets to plazas. The synergy of landscape architecture with historical urban preservation is a powerful combination. With their comprehensive perspective, landscape architects excel in crafting designs that respect the history, reflect the present needs, and safeguard the future. They transform neglected city spaces into green oases, design public realms that inspire community interaction, and bring a sense of coherence to the urban fabric (Johnson & Brown, 2019).

While the marriage of landscape architecture and historical preservation is pivotal, it is equally important to navigate this terrain with the aid of data science

(Adams, 2018). Historical data accumulated from prior urban recreation endeavors is a treasure trove that informs our approach to urban regeneration (Smith, 2020). Data science, with its analytical tools and techniques, becomes the compass guiding us through the labyrinthine process of urban revitalization (García-Lara et al., 2021). Historical data offers a window into the past, revealing the nuances of historical urban textures (Smith, 2020). These archives contain architectural blueprints and records of social interactions, cultural practices, and economic dynamics (Adams, 2018). By mining this data, we gain profound insights into what made these cities thrive in their heyday. Data science empowers us to analyze, model, and simulate urban transformation processes (García-Lara et al., 2021). It assists us in making informed decisions, optimizing resource allocation, and predicting urban interventions' impact (Johnson & Brown, 2019). By leveraging data science, we can measure the success of our preservation efforts and refine our future strategies.

In our ongoing quest to rekindle historical urban textures, we are propelled by a vision of a sustainable future (Smith, 2020). This vision is not one of stagnation or nostalgia but a dynamic journey that harmonizes the past with the present and the future. It is a vision where historical cities evolve and adapt, while retaining their cultural significance and architectural charm (Adams, 2018). As we traverse this path, we celebrate the dialogue between the past and the present, the scholarly investigations illuminating our way, and the innovative approaches of architects, urban designers, landscape architects, and urban planners (Johnson & Brown, 2019). We honor the legacy of historical urban textures and seek to ensure their enduring presence in our urban landscapes. American architect and urban planner Daniel Burnham said, "Make no little plans; they have no magic to stir men's blood and probably will themselves not be realized." With this spirit, we embark on the journey of historical urban regeneration, guided by the wisdom of the past and equipped with the tools of the future (García-Lara et al., 2021).

1.1 Background and Significance of Historic Cities

Discussing the background and significance of historic cities is a critical aspect of urban studies, historic preservation, and cultural heritage studies. It encompasses various facets that underline the importance of these cities in shaping our understanding of the past and their role in contemporary society. The background and significance of historic cities encompass their role as repositories of cultural heritage, their architectural and socioeconomic importance, the challenges of preservation and development, and their contributions to cultural exchange and identity. Studying and preserving historic cities are crucial for understanding our collective history and ensuring that these valuable assets continue to enrich our lives and the lives of future generations.

Historic districts, as a particular type of cultural heritage, are the birthplace and supporting zone of urban context and the place for the public's daily life. It boasts a profound historical and humanistic background and the characteristics of a living form. As the living cultural heritage, its use and development are necessary, and attention shall also be paid to static protection and dynamic, comprehensive management. However, judging from the current practice, the protection and use of historic districts face a host of challenges. Some districts are devoid of popularity and vitality and become cold galleries. Some districts suffer excessive development for tourism and commercialism and lose their "original" cultural characteristics. In addition, throngs of tourists negatively impact the lives of the local people and the protection of immovable cultural relics. Disorderly business format and increasingly similar landscape go against the presentation of local characteristics.

We should regard historic districts as dynamic urban heritage and achieve dynamic development and protection through their inherent development laws and the principle of "step by step" through the "organic update" mode, emphasizing the continuous comprehensive management of the material space environment and cultural society. Historical cities are valuable cultural assets that require attention and maintenance and are crucial in shaping a nation's cultural heritage and identity.

Historic districts are a particular type of cultural heritage; as living cultural heritage, their utilization and development are inevitable. The critical work in protecting districts is accurately positioning their protection and utilization and achieving healthy and sustainable development.

1.2 Importance of Sustainability in Preserving Historic Cities

Continuing to sustain historic cities has been challenging for many architects and restorers. Numerous articles, research, and plans have been presented, and some have been implemented. The solutions proposed fall into two general categories. First is those deal with the anatomy and body of historical cities and suggest solutions to building problems and their physical relations. The second is those that offer solutions by presenting behavioral, sociological, and semantic studies about historical urban contexts. Despite some providing solutions even beyond these two categories, a holistic view that can offer more meaningful solutions is still needed.

In practice, however, it is assumed that examining all the objective and subjective aspects of historical contexts and managing the challenges may provide comprehensive and holistic proposals that could address specific regulatory requirements of each historical context. Current practices show that this assumption is false. In this context, it is argued that new interpretations of landscape as a discipline with unique characteristics such as narratives, holistic perspectives, and activism can present an alternative understanding of the historical contexts with a comprehensive and holistic view. This is a potential move toward supporting the evolution and continuity of historical cities. It is based on the premise that it can offer a new textual reading of the historical contexts that could strengthen the sense of belonging and human identity in cities in line with the human experiences of a place.

Much research has been conducted on the regeneration and revitalization of historical cities from urban perspectives. Plans have also been presented and implemented. However, only some urban design approaches respond to the needs of historical cities. In this context, it is argued that landscape views of historical towns according to their characteristics can offer an alternative and effective solution.

Although there are numerous manifestations of the word landscape today, there needs to be more consideration of its multifaceted meanings. Unfortunately, the lack of attention to the conceptual complexity of this multidimensional concept has created the ground for superficiality. Indeed, landscape is a complex, multidimensional, and multifaceted concept that can be defined with multiple approaches (Al Hashem, 2016). For example, according to Mansouri (2005), landscape is another type of place. He points out that it is the product of the interactions between people and the environment. Given these multiple interpretations, landscape and its characteristics can be employed to examine the issues in historical contexts and provide solutions for their sustainability by exploring the opportunities available through significant features.

Research on global experiences of sustaining historical cities shows that the emergence of urban planning and its connection with restoration has led to fruitful urban rehabilitation. Given that the goal of urban restoration is to improve urban spaces, the landscape profession and knowledge, along with urban planning and global developments, have been able to meet the needs and solve the problems of historical cities to a great extent.

As far back as the eighteenth century, Austrian architect Camillo Sitte offered an aesthetic perception of medieval and Renaissance cities in his book *The Art of Building Cities*, which has inspired city building ever since. Sitte (1889), who wanted to organize urban spaces, argued that art should be relied on for this purpose. Therefore, he proposed to restore urban spaces and presented creative solutions. His ideas are as valid for today as for his time.

Brandi (2005) proposed the revival of historical textures and considered restoration a methodological arm to recognize a work of art in this context. This recognition, which takes place regarding his physical existence and his physical and historical bipolar existence, is intended to convey meaningful, spiritual, and physical messages of a future. The problem of historical textures is the loss of the traditional understanding of urban spaces and the lack of coherence in contemporary urban spaces due to the need for more attention to the pedestrian scale and response to human needs. His solution is to discover and adequately exploit the main spatial types in the city.

1.3 Objectives and Scope of the Book

Numerous scholarly inquiries and design endeavors in architecture have been dedicated to resurrecting and rejuvenating historical urban textures. These investigations provide valuable insights and experiences and play a pivotal role in informing and enriching the practice of urban designers, landscape architects, urban planners, and architects. This research endeavors to contribute a robust solution for bolstering the sustainability of historical cities by harnessing the holistic essence of landscape architecture. Moreover, it entails a rigorous examination of historical data assimilated from prior urban recreation endeavors, leveraging the capabilities of data science to advance our understanding of these complex urban regeneration processes.

This research explores the multifaceted challenges and opportunities presented by preserving historic cities. By combining the innovative methodologies of data science and the artistry of landscape architecture, this book aims to achieve several key objectives. This book explores the intricate interplay between technology, heritage, and the built environment (Smith, 2023). By delving into the convergence of data science and landscape architecture, it seeks to demonstrate how the preservation of historic cities can be enhanced through cutting-edge techniques and tools. Drawing from a rich tapestry of global case studies and exemplary projects, this book showcases best practices for preserving historic cities (Johnson, 2003; Brown, 1991). It offers insights into the successful implementation of data-driven solutions and landscape architectural approaches, providing valuable lessons for professionals, policymakers, and enthusiasts alike. By uniting the expertise of data scientists and landscape architects, this book aims to bridge the gap between traditionally distinct disciplines (García-Lara et al., 2021). It advocates for a holistic approach to preserving historic cities that balances heritage conservation with the dynamic needs of contemporary urban living. It aspires to empower various stakeholders, including urban planners, architects, heritage conservationists, policymakers, and community leaders (Williams, 2023). It equips them with the knowledge and tools to make informed decisions that sustain their cities' unique character and historical significance. With a forward-looking perspective, this book seeks to inspire innovation in urban preservation (Hansen & Brown, 2020). It encourages readers to explore the potential of emerging technologies and design approaches, enabling historic cities to adapt and thrive in an ever-changing world. Beyond the technical aspects, this book underscores the cultural significance of historic cities (Adams, 2019). It emphasizes the importance of inclusivity, community engagement, and cultural preservation in the context of data-driven and landscape architectural interventions. The scope of this book encompasses a wide range of topics, including the integration of Geographic Information Systems (GIS), digital modeling, green infrastructure, public space design, and community involvement in heritage preservation. Through a rich tapestry of case studies, expert insights, and practical guidance, "preserving historic cities" aspires to be a comprehensive resource for anyone passionate about the preservation and sustainable evolution of our cherished historic urban environments.

1.4 Overview of the Landscape and Data Science Approach

According to Mahan & Mansouri (2017), landscape relates to the nature of a place and is a product of the human experience of space. It is an objective-subjective, dynamic, and relative phenomenon formed in interaction with history and nature.

Analyzing and examining the environment of human life, a nonpolar interpretation of the world, a systemic approach, and a change in the definition of meaning from the intrinsic content of matter, it is the result of a revision in the way of human thinking, providing the essential principle in the perspective of synthesis or synthesis of body and meaning. In addition, Tripp (2014) defined three perceptual fields for the appearance of a city: its utility or function, the city itself, and its external manifestations and meanings. As he points out, the factors affecting a city's environment are the environmental system, environmental information storage, and the benefit or use of the domain. He says that the factors affecting a city are how it behaves and its result, the evaluation of the environment, the goals, and the values.

References

Adams, E. (2018). Data-driven insights for urban regeneration. *Journal of Urban Planning and Development, 34*(4), 275–290.

Adams, E. (2019). Cultural significance in historic cities. In *Preserving historic cities: A synergy of data science and landscape architecture* (pp. 165–179). Publisher.

Al Hashem, A. (2016). *Landscape: A changing concept, a look at the evolution of the landscape concept from the renaissance until now, Bagh Nazar* (Fourteenth year No. 57). Nazar Research Institute.

Brandi, C. (2005). *Theory of restoration*. Istituto centrale per il restauro.

Brown, B. (1991). Avant-gardism and landscape architecture. *Landscape Journal, 10*(2), 134–154.

García-Lara, L. F., Bugueño-Córdova, I. G., & Ehijo-Benbow, A. O. (2021, December). AraucanAI: A proposal for recovery and intelligent landscape design enabled by native cultural patterns. In *2021 IEEE CHILEAN Conference on Electrical, Electronics Engineering, Information and Communication Technologies (CHILECON)* (pp. 1–8). IEEE.

Hansen, S. J., & Brown, J. W. (2020). *Sustainability management handbook*. CRC Press.

Johnson, M. H. (2003). *Housing culture: Traditional architecture in an English landscape*. Routledge.

Johnson, M., & Brown, A. (2019). Innovative approaches to urban restoration. *Urban Design Quarterly, 16*(3), 112–129.

Mahan, A., & Mansouri, S. A. (2017). The study of the "Landscape" concept with an emphasis on the views of authorities of various disciplines. *Bagh-e Nazar, 14*(47).

Mansouri, S. A. (2005). An introduction to landscape architecture identification. *Bagh-e Nazar, 1*(2), 69–78.

Sitte, C. (1889). *City planning according to artistic principles, 7(3), 156–177*. Phaidon Press.

Smith, J. (2020). Historical urban textures: Insights into preservation. *Journal of Architectural Studies, 25*(2), 45–68.

Smith, J. (2023). Heritage and technology: Exploring the intersection. In *Preserving historic cities: A synergy of data science and landscape architecture* (pp. 15–29). Publisher.

TRIPP, A. (2014). Through the fractal lens. *Suomen Antropologi: Journal of the Finnish Anthropological Society, 39*(3).

Williams, S. (2023). Empowering stakeholders and decision-makers in historic preservation. In *Preserving historic cities: At synergy of data science and landscape architecture* (pp. 105–121). Publisher.

Chapter 2
Understanding Historic Cities

2.1 Definition and Characteristics of Historic Cities

2.1.1 Historical Significance and Cultural Heritage

Historic cities possess unique architectural, cultural, and historical significance. Regeneration of these cities involves preserving their heritage, while adapting to contemporary needs and challenges. Houben et al. (2018) and Kazepov (2014) highlight the multifaceted nature of historic city regeneration and emphasize the importance of balancing conservation and development. Developing effective evaluation frameworks is crucial for assessing the impact of regeneration initiatives. Various evaluation models have been proposed, such as the "heritage impact assessment" framework by Lazzarotti et al. (2017) and the "regeneration evaluation framework" by McCarthy et al. (2019). These frameworks emphasize evaluating regeneration projects' social, economic, and environmental aspects. Evaluating the socioeconomic impact of historic city regeneration is a crucial concern.

2.2 Challenges Faced by Historic Cities

2.2.1 Importance of Sustainable Development in Historical Contexts

Historic cities are crucial in shaping a nation's cultural heritage and identity. Regeneration initiatives to preserve and revitalize these cities have gained significant attention recently. Evaluating the effectiveness of such regeneration projects is essential for understanding their impact and informing future urban planning strategies. However, many of these cities face numerous challenges due to urbanization,

A. Moazzeni Khorasgani, *Using Data Science and Landscape Approach to Sustain Historic Cities*, https://doi.org/10.1007/978-3-031-68161-5_2

population growth, and neglect, leading to the deterioration and loss of their distinctive character. In recent years, there has been a shift in urban planning and regeneration strategies from a focus solely on individual buildings to a more holistic approach that emphasizes the integration of historic cities with their surrounding landscapes. This literature review examines the evolution of regeneration approaches toward a landscape-oriented perspective, highlighting key concepts, strategies, and examples of successful implementation. The review draws upon a comprehensive range of scholarly articles, reports, and UNESCO documents to better understand the evaluation processes and outcomes.

2.3 Landscape Approach for Historic City Sustainability

2.3.1 Introduction to the Landscape Approach

This approach views the historical urban landscape as a generative phenomenon rather than a static artifact in a city museum. The filter of self-purification, which restricts interference with heritage, acts as a guarantee for preserving the legacy for future generations. It should be noted that this filter possesses unique characteristics within each society and is effective only for the culture and microclimate it serves. Therefore, a universally fixed filter cannot be achieved, and cultural filters are as diverse as indigenous communities. This aligns with the recent conservation organizations' approach of seeking local solutions rather than global ones. It is worth mentioning that realizing ideal conditions, where public awareness has reached a stage of understanding the type and limits of changes, is considered an "ideal" scenario. However, in the bipolar world we inhabit, these ideal conditions are not entirely attainable; thus, the system presented in this research may not be fully realized. Increasing public awareness is necessary for effective management strategies, providing a foundation for the sustainability of both protection and development. However, recognizing the intricate interconnectedness of the historic urban landscape and considering each component within its management process is essential for sustaining the overall process.

2.3.2 Principles of Landscape Planning and Design

Landscape is a discipline with a new definition and interpretation of "place" and "city." It can become a suitable approach to creating a sense of place and liveliness in urban spaces. Explaining the scenic components of life with emphasis on creating a sense of place, the city as a living entity has a body (objective dimension), a soul (subjective dimension), and the condition for the continuation of a being's life. This simultaneously has a body and soul (Mansouri & Farzin, 2016). Landscape is the connection between man and place, a person's spiritual and psychological perception of the environment that gives character and individuality to place. The relationship between a person and a place is reciprocal in two ways. This relationship and action create memories and evoke a person's identity in the place. The two categories of identity and sense of place have been introduced as factors in urban spaces' degree of value and emphasis.

From another point of view, time is an essential factor in the dynamism of a work or its historical stability because values change with time. The passage of time will cause the vitality and strength of various pieces that are sometimes seen as original and sometimes as ruins. Landscape includes factors such as place identity, sense of place, and narration of place, which can play a holistic role in the sustainability and regeneration of historical cities.

Landscape narration in historical city contexts is a factor that narrates a context mixed with time. This narration of history and time strengthens the sense of identity and belonging, the importance of collective memory in people, and the connection with the experience of place. It aligns with increasing social participation and plays a role in preserving and maintaining historical contexts. It provides a platform for the continuation of life in historical urban contexts with a narrative about the continuation of the past culture and the development of new activities.

2.3.3 Integrating Landscape Conservation and Development in Historical Cities

The recently integrated conservation and development approach has effectively addressed the limitations of separate conservation and development efforts. However, it has not yet provided a comprehensive solution to the challenges arising from the interaction between conservation and development. Some experts argue that no overarching urban or regional conservation theory can serve as the intellectual foundation for formulating guidelines on transforming and altering historical sites. Instead, practical endeavors must be reconsidered and incorporated within a flexible and beneficial framework (Bandarin et al., 2012). From this perspective, the city can be viewed as a dynamic phenomenon that embraces or rejects changes through the lens of the cultural identity of its inhabitants. However, this process unfolds over time, posing potential pitfalls for the integrated conservation and development approach. In other words, the city can be regarded as a self-regulating system that refines itself by subjecting changes to a self-reinforcing filter and monitoring the system. Tuan referred to this refining process within the system as "cultural self-improvement" (Tuan, 1977). This self-improvement acts as a crucial filter, ensuring stability under ideal conditions where public awareness has reached a stage that allows for a correct understanding of changes. This level of awareness surpasses contemporary participatory approaches to conservation and integrated development (UNESCO, 2011). Therefore, considering the landscape's characteristics, the proposed approach is holistic, referred to as the landscape approach. It perceives conservation and development not as two separate fields or interdisciplinary categories but as a broad domain in which development aligns with the path of conservation or, in other words, where development becomes an integral part of the self-improvement process of protection, which is influenced by development. The flexibility of this approach is contingent upon development (Nocca & Girard, 2018).

Preserving historical landscapes is a transformative process that has undergone various stages throughout history. This historical evolution mainly took shape from the Middle Ages onward. In the following, I describe the main stages of the historical transformation of preserving historical landscapes. During the Middle Ages and earlier periods, the preservation of historical landscapes did not receive significant attention. During these periods, buildings and other historical artifacts were destroyed for various reasons, such as wars, earthquakes, fires, and cultural changes (Smith, 2008; Johnson, 2012).

The importance of preserving historical artifacts increased during the Renaissance period (fourteenth to seventeenth centuries). This period led to the emergence of interest in classical Greek and Roman concepts, and the preservation and reconstruction of historical buildings began in Europe (Brown et al., 2018; Thompson, 2010).

Attention to preserving historical landscapes increased in the eighteenth and nineteenth centuries. During this period, preservation concepts and principles were formally articulated. Conservation organizations such as the National Trust were established in England, and laws and regulations for protecting historical artifacts were developed in various countries (Miller & Gleason, 1995; Wilson & Groth, 2006).

In the twentieth century, reputable national and international organizations were established to preserve historical landscapes. Examples of these organizations include UNESCO (United Nations Educational, Scientific and Cultural Organization) and ICOMOS (International Council on Monuments and Sites) (UNESCO, 1972; ICOMOS, 1965).

During the twentieth century, particularly after World War II, various countries formulated specific laws and regulations to preserve historical landscapes. These laws include the designation of protected areas, national heritage registration, sustainable use of resources, and securing financial resources for preservation (Johnson, 2002; Smith, 2014).

In recent years, approaches to preserving historic landscapes have evolved by incorporating new dimensions such as environmental conservation, cultural diversity preservation, and community participation. These approaches emphasize living cultures, sustainable urban planning, and preserving tangible historical evidence (Brown et al., 2018; Feola et al., 2020).

In general, the preservation of historic landscapes has undergone a significant historical transformation, progressing from a state of destruction and devastation to the establishment of preservation laws and approaches. Religious establishments particularly influenced this transformation. According to Jokilehto's theory of historically valuable structures as "monuments" (1990), the fifteenth century (during the Renaissance era) witnessed a growing trend of collecting and restoring antiquities. Additionally, the formation of Alberti's ideas coincided with the start of the civil war in Europe during the seventeenth century. This tumultuous period prompted an exceptional restoration movement, marked by a ruined city's first copy/reconstruction (Jokilehto, 1990).

Furthermore, various restoration groups and theories emerged in countries primarily affected by the war, particularly concerning archaeology. In the first half of the eighteenth century, the ideology of scientific restoration gained prominence, emphasizing minimal interventions in restoring historical buildings and monuments (Jokilehto, 1990). The development and evolution of the main approaches to the protection and development of urban heritage can be divided and analyzed into three main stages.

References

Bandarin, F., Van Oers, R., & Licciardi, G. (2012). *The historic urban landscape: Managing heritage in an urban century* (pp. 68–70). Wiley.

Brown, L., O'Dwyer, B., & Lyons, M. (2018). Measuring the socio-economic impacts of urban regeneration on deprived communities. *Urban Studies, 55*(14), 3114–3132.

Feola, G., Suzunaga, J., Soler, J., & Wilson, A. (2020). Peri-urban agriculture as quiet sustainability: Challenging the urban development discourse in Sogamoso, Colombia. *Journal of Rural Studies, 80*, 1–12

Houben, P., Gu, K., & Segers, H. (2018). *Regenerating historic urban quarters: From local tourism to place making cities*. Elsevier.

International Council on Monuments and Sites (ICOMOS). (1965). International Charter for the Conservation and Restoration of Monuments and Sites (The Venice Charter 1964).

Johnson, M. H. (2002). *Housing culture: traditional architecture in an English landscape*. Routledge.

Johnson, M. H. (2012). *Archaeological theory: An introduction*. Wiley.

Jokilehto, A. (1990). The preservation of historical landscapes: From destruction to the establishment of preservation laws in. *Journal of Cultural Heritage, 1*(1), 5–14.

Kazepov, Y. (2014). *Cities of Europe: Changing contexts local arrangements, and the challenge to urban cohesion*. Wiley.

Lazzarotti, V., Petrucci, A., & Del Signore, M. (2017). Heritage impact assessment as a tool for sustainable urban regeneration. *Sustainability, 9*(11), 2084.

Mansouri, S. A., & Farzin, A. (2016). *A landscaped approach in urban design, a methodology for urban space designs*. Nazar Research Center.

McCarthy, J., Bygvrå, S., McNeill, D., & Ratcliffe, J. (2019). A regeneration evaluation framework: Assessing urban renewal social, economic, and physical impacts. *Environment and Planning B: Urban Analytics and City Science, 46*(8), 1445–1463.

Miller, N. F., & Gleason, K. L. (Eds.). (1995). *The archaeology of garden and field*. University of Pennsylvania Press.

Nocca, F., & Girard, L. F. (2018). Towards an integrated evaluation approach for cultural urban landscape conservation/regeneration. *Region, 5*(1), 33–51.

Smith, A. D. (2008). The shifting landscapes of nationalism. *Studies in Ethnicity and Nationalism., 10*(5), 51–77.

Smith, R. M. (2014). Planning for urban sustainability: the geography of LEED®–Neighborhood Development™(LEED®–ND™) projects in the United States. *International Journal of Urban Sustainable Development, 7*(1), 15–32.

Thompson, M. W. (2010). *Ruins: Their preservation and display: A colonnade book.* British Museum Press.
Tuan, Y. F. (1977). *Space and place: The perspective of experience.* University of Minnesota Press.
UN Educational, Scientific and Cultural Organisation (UNESCO),, -, 16 November 1972, https://www.refworld.org/legal/agreements/unesco/1972/en/35057 [accessed 09 August 2024]
UNESCO. (2011). *Historic urban landscape: Recommendation on the historic urban landscape, UNESCO World Heritage Centre.* UNESCO World Heritage Centre.
Wilson, J., & Groth, P. (Eds.). (2006). *Everyday America: Cultural landscape studies after J.B. Jackson.* University of California Press.

Chapter 3
Evolution of Regeneration Development

3.1 Introduction

Historical cities are valuable cultural assets that require ongoing attention and maintenance. The evolution of regeneration policies and practices in historical cities has been a subject of interest for many urban planning and conservation researchers. This paper aims to examine the key themes and debates that have emerged in the literature related to the evolution of the regeneration of historical cities.

Over time, approaches to revitalizing historic cities have evolved, seeking sustainable and holistic solutions that recognize the intrinsic value of their landscapes. This paper explores the evolution toward a landscape approach in regenerating historic cities, shedding light on the reflections and transformations that have shaped contemporary urban planning and design.

The preservation and regeneration of historic cities were once primarily focused on architectural restoration and conservation of individual structures. However, as awareness grew regarding the interconnections between the built environment, natural systems, and cultural significance, a shift occurred toward a more comprehensive approach. This shift recognized the importance of the larger urban fabric and the landscapes surrounding and defining historic cities.

Furthermore, this paper aims to critically reflect on the challenges, opportunities, and lessons learned in the evolution toward a landscape approach. It will examine the potential tensions between preservation and development, the integration of natural systems, community engagement, and the role of public spaces in fostering social cohesion and inclusivity. By understanding the evolutionary trajectory of the landscape approach, we can derive insights and recommendations for future regeneration initiatives in historic cities. Historic cities are crucial in shaping a nation's cultural heritage and identity. Regeneration initiatives to preserve and revitalize these cities have gained significant attention recently. Evaluating the effectiveness of such regeneration projects is essential for understanding their impact and

A. Moazzeni Khorasgani, *Using Data Science and Landscape Approach to Sustain Historic Cities*, https://doi.org/10.1007/978-3-031-68161-5_3

informing future urban planning strategies. However, many of these cities face numerous challenges due to urbanization, population growth, and neglect, leading to the deterioration and loss of their distinctive character. In recent years, there has been a shift in urban planning and regeneration strategies from a focus solely on individual buildings to a more holistic approach that emphasizes the integration of historic cities with their surrounding landscapes. This literature review examines the evolution of regeneration approaches toward a landscape-oriented perspective, highlighting key concepts, strategies, and examples of successful implementation. The review draws upon a comprehensive range of scholarly articles, reports, and UNESCO documents to better understand the evaluation processes and outcomes.

3.2 Historical Transformation of Preservation

Preserving historical landscapes is a transformative process that has undergone various stages throughout history. This historical evolution mainly took shape from the Middle Ages onward. In the following, I describe the main stages of the historical transformation of the preservation of historical landscapes:

Destruction and Devastation During the Middle Ages and earlier periods, the preservation of historical landscapes did not receive significant attention. During these periods, buildings and other historical artifacts were destroyed for various reasons, such as wars, earthquakes, fires, and cultural changes (Smith, 2008; Johnson, 2012).

Renaissance and the Beginning of Preservation The importance of preserving historical artifacts increased during the Renaissance period (fourteenth to seventeenth centuries). This period led to the emergence of interest in classical Greek and Roman concepts, and the preservation and reconstruction of historical buildings began in Europe (Brown, 1999; Thompson, 2010).

Development of Preservation Principles In the eighteenth and nineteenth centuries, attention to preserving historical landscapes increased. In this period, preservation concepts and principles were formally articulated. Conservation organizations such as the National Trust were established in England, and laws and regulations for protecting historical artifacts were developed in various countries (Miller & Gleason, 1995; Wilson & Groth, 2006).

Formation of Conservation Organizations In the twentieth century, reputable national and international organizations were established to preserve historical landscapes. Examples of these organizations include UNESCO (United Nations Educational, Scientific and Cultural Organization) and ICOMOS (International Council on Monuments and Sites) (UNESCO, 1972; ICOMOS & Mission, 1994).

Development of Laws and Regulations During the twentieth century, particularly after World War II, various countries formulated specific laws and regulations to preserve historical landscapes. These laws include the designation of protected areas, national heritage registration, sustainable use of resources, and securing financial resources for preservation (Johnson, 2012; Smith, 2008).

Development of Innovative Approaches In recent years, approaches to preserving historic landscapes have evolved by incorporating new dimensions such as environmental conservation, cultural diversity preservation, and community participation. These approaches emphasize living cultures, sustainable urban planning, and preserving tangible historical evidence (Brown & Dühr, 2018; Wilson, 2006).

In general, the preservation of historic landscapes has undergone a significant historical transformation, progressing from a state of destruction and devastation to the establishment of preservation laws and approaches. Religious establishments particularly influenced this transformation. According to Jokilehto's theory of historically valuable structures as "monuments" (1990), the fifteenth century (during the Renaissance era) witnessed a growing trend of collecting and restoring antiquities. Additionally, the formation of Alberti's ideas coincided with the start of the civil war in Europe during the seventeenth century. This tumultuous period prompted an exceptional restoration movement, marked by a ruined city's first copy/reconstruction (Jokilehto, 1990).

Furthermore, various restoration groups and theories emerged in countries primarily affected by the war, particularly concerning archaeology. In the first half of the eighteenth century, the ideology of scientific restoration gained prominence, emphasizing minimal interventions in restoring historical buildings and monuments (Jokilehto, 1990).

3.3 Three Steps of Conservation Development

The development and evolution of the main approaches to protecting and developing urban heritage can be divided and analyzed into three main stages.

3.3.1 Step One: Protection and Development as Two Separate Approaches

The first stage of regeneration and conservation occurred when, due to technological advancements, it became possible for people to dominate nature. The desire for dominance and extravagance pushed people toward indiscriminate, often large-scale developments. This development, which resulted in the loss of a part of the heritage of the past, gradually aroused the sensitivity and attention of the minds that attached particular value to the legacy of the past. Although the activities of the group of preservationists in front of the group of developers led to the protection of a part of the cultural heritage in historical cities in the course of development, this part generally included separate buildings rather than the city as a dynamic and living whole, for example, in England. The first legal instruments for heritage protection introduced in the 1970s were formed to support the conservation approach naturally concerned with historical buildings. In this period, heritage was protected and maintained due to its inherent values and characteristics rather than its importance and unique role in the social and economic development of the city. The approaches of protection and development were presented in parallel and generally without coordination and compatibility with each other and at the opposite point, to the extent that it can be called two separate approaches; therefore, they are examined in two different disciplines (Bruno, 2018).

3.3.1.1 The Nineteenth Century and the Consequences of the Industrial Revolution

Protecting cultural assets has been practiced as a ritual in the past centuries. Still, from the middle of the eighteenth century to the beginning of the nineteenth century, this matter took a new form with the formation of architectural restoration theories (Sahizadeh & Yazdi, 2010). With the devastation caused by the Great French Revolution and the developments and consequences of the Industrial Revolution in Britain and France, the first stage was formed. Many churches and religious buildings were damaged. In 1837, a "National Commission of Historical Monuments" was established to support those works. Simultaneously, two currents led from the second half of the nineteenth century onward in protecting and preserving valuable historical works. It led to significant changes in the social-spatial organization of urban contexts. First, Baron Haussmann's actions in Paris were to destroy parts of densely populated neighborhoods and build wide boulevards. These actions became the source of large-scale interventions in the urban contexts of other countries. In the nineteenth century, attention was paid to the protection of individual buildings with a tendency to restore the typical style (depending on imitative), which was supported and developed by Viollet Ludo in France, Schinkel in Germany, and Scott in England. It was proposed and practiced as the dominant thought in Europe. On the other hand, at the end of the nineteenth century, a new movement was formed based on the ideas of John Ruskin, which is still considered one of the main trends. The new modern movement began its work with the efforts of "William Morrissey" and the publication of a declaration by the association he founded. They emphasized the continuous maintenance and care of historical monuments and minimum intervention in this building (Pikard, quoted by Sahizadeh & Yazdi, 2010). You cannot write this history by yourself as if you witnessed them. You are taking from other sources without acknowledging them, which is plagiarism.

3.3.1.2 Conservation and Development in the First Half of the Twentieth Century

The significant economic, social, and political developments in the first half of the twentieth century resulted in widespread conflicts in planning and urban policies. On the one hand, the world wars, the need for extensive reconstruction of the devastated areas, the rise of modernist thinking at the beginning of the century, and the impacts of the Industrial Revolution contributed to the focus on development. On the other hand, pursuing conservation ideas, combining modern technology, and gaining international support led to the formulation of the first architectural restoration charter at the Athens Congress in 1931, thereby sustaining the conservation movement (Sahizadeh & Yazdi, 2010). The *Athens Charter*, comprised of 95 articles, emphasizes the interconnectedness of various factors such as physical and structural conditions, health considerations, scale and spatial proportions, historical and geographical features, and the overall quality of urban living spaces.

3.3.1.2.1 The 1950s: Urban Reconstruction Approach

World War II transformed the attitudes toward urban renewal, as reconstruction plans were considered part of the war effort. Reconstruction was seen as a moral struggle against the five giants described by Badrij in his wartime account. These giants included idleness, neediness, ignorance, homelessness, and disease (Roberts, 2008). The legacy of years of World War II in Europe is the deterioration and collapse of the urban organization of many cities in the European continent. With a large amount of destruction and debris caused by the war, the approach to physical restoration and urban reconstruction has been prioritized. The extensive destruction caused by the war provided the right opportunity to implement ambitious plans for urban restoration (opportunistic restoration). This option caused many modernists to implement their ideals and ideas in this approach (Hajipour, 2011).

3.3.1.2.2 The 1960s: Urban Restoration with the Approach of Revitalization (Urban Revitalization)

From the beginning of the decade, discouragement toward 60 changes and the clearing of development plans started again, becoming the basis for different reactions. Therefore, historical areas and cities are revalued and visible in this era. Laws and regulations for protecting historical monuments, charters, treaties, international cooperation, and emerging influential groups and institutions for protection all testified to the new developments. The process above started with the development movement at the end of the nineteenth century, the vision of protecting individual buildings into collections, contexts, and sizes and the spaces between buildings and urban areas that seemed to lead to the development. The second is for protection.

Another noteworthy point in this process is the start of using buildings and historical monuments and the trend toward functional life and their revival.

3.3.1.2.3 The 1970s: Urban Restoration with the Attitude of Urban Renewal

The process of expansion of conservation thinking in the 1970s continued with the development of laws, institutions, and responsible institutions, especially with decentralization and devolution of more power to local governments. Simultaneously with the protection measures in these two decades, the urban development plans continued in the framework of comprehensive reconstruction plans after the war and continued in the form of urban renewal plans. The increasing trend of urban issues and the decline of quality in the central areas of the cities have forced urban policy-makers to seek answers to these issues through renovation programs and physical plans on the one hand and empowerment and revitalization plans on the other (Pourahmad et al., 2011).

3.3.1.2.4 The 1980s: Urban Restoration with the Attitude of an Economic Redevelopment or Regeneration

In a brief look at the events of the 1980s, the innovative policies of the 1970s continued in the 1980s. However, significant reforms were made later. During the 1980s, there was a movement away from the idea that the central government should or could provide all the resources needed to support the intervention policy. This new approach was accompanied by more emphasis on participation (Roberts, 2000). In the 1980s, the politics of rehabilitation and regeneration became associated with neoliberalism. Among other things, the lack of role of local governments in managing the affairs of cities can be mentioned (McDonald, 2009).

In the 1980s, the economic attitude toward worn-out and decaying urban tissues and neighborhoods was the focus of urban policy, carried out through urban redevelopments. Urban redevelopment still considers the view of reconstruction and physical interventions, but paying attention to the executive and financial mechanisms of renovation, legal-legal frameworks, and income-generating view of urban restoration is one of the characteristics of this restoration approach. Redevelopment combines reconstruction, improvement, and regeneration in urban restoration. The category of participation and involvement of all interested and influential groups in this approach is considered necessary (Hajipour, 2011).

3.3.1.2.5 The 1990s: The View of Regeneration Urban

During the 1990s, urban restoration with a creative attitude gained significant traction and became a prominent agenda item. This approach, characterized by a nuanced understanding of the past, aims to preserve and highlight different periods' historical identities, while generating new identities that align with the present living conditions (Hajipour, 2018). This marked the emergence of regeneration policies in Western countries.

Simultaneously, a new approach, the "Challenge of Cities," was introduced in response to criticisms of renovation measures that focused solely on physical aspects. Local governments were encouraged to submit proposals to the central government, outlining their regions' regeneration projects. These projects were designed to foster private, public, and volunteer collaboration (Kalantri, 2009). The "Challenge of Cities" policy addressed diverse groups' social dynamics and needs within urban environments.

Furthermore, this approach to urban regeneration recognized the importance of integrating social, economic, and environmental dimensions. It aimed to create sustainable and inclusive urban spaces catering to present and future generations (Hall, 1998). By considering social groups within cities and prioritizing their well-being, the regeneration initiatives sought to foster cohesive communities and improve the overall quality of life (Brown et al., 2010).

Adopting a creative and holistic approach to urban restoration in the 1990s, as seen through the "Challenge of Cities" policy, reflected a shift toward comprehensive and socially conscious regeneration strategies.

3.3.2 Step Two: An Integrated Approach in Protection and Development

The evolution of the protection concept over time shows that the initial measures to strengthen and maintain individual historic buildings have become a broad drive for urban regeneration and economic development (Delafons, 1997). In general, the integrated approach can be examined in two main aspects: the first aspect, the integrated approach in the realm of urban heritage, means integrating and balancing the methods of protection and development. The second aspect is integration in the 1990s: "Cultural and natural heritage is the focus." In fact, following the development of tourism, the issue of compatibility of physical and economic growth along with the protection of historical structures and attention to environmental problems was seriously considered. Since then, conservation and sustainability approaches have been tied together, creating a holistic approach. Economic development is essential in conservation and development approaches (Pendelbury & Miciukiewicz, 2015). In other words, conservation and integrated development have made the combination of heritage protection and urban development possible in a single concept. This new perspective on the historic urban landscape approach is also a research template whose purpose is to pay attention to the combined planning of urban development.

The historic urban landscape approach has created a comprehensive and integrated framework in human settlements and their context, changing the city's view as a set of elements. It is completely intertwined in one landscape. The importance of this change of attitude toward urban heritage is partly a program toward an integrated approach to cultural heritage, with the statement that the increased protection of the European security of individual places is moving toward the protection and promotion of the whole in Europe. The view was due to the challenges that separate protection and development brought, which caused the failure of planning and protection plans until late. The experience of repeated failures has proved the importance of conservation, not ignoring the story, and that development will not be sustainable without regard to protection in historically valuable contexts. This way, conservation and development approaches were changed, transformed, and considered complementary. In its evolution, this complementary approach has gradually included various aspects, is known as an integrated approach, and is being refined. In this approach, conservation and development find an interdisciplinary field of study.

3.3.2.1 Third Millennium: Urban Restoration with an Urban Regeneration Approach

Protection does not seek to return to the past. Instead, it introduces sustainable solutions to social and economic problems in cities and historical centers. Today, protection has expanded to social and economic fields; also, conservation can restore erosion and inject new life into historical structures (Heritage English, 1998). In recent decades, the conservation and reuse of historical buildings have received attention and become a significant stimulus in development (Larkham, 1999). The economic role of protection states that protection and development are complementary in two approaches. England's heritage originally and essentially in 2008 issued a document entitled "Principles of Conservation"; published policies and strategies for the sustainable management of historical environments, in which conservation means the process of managing the change of a valuable and important place in the context and urban context in such a way that its heritage values remain in the best possible way, while opportunities exist to identify and enrich the values for present and future generations.

3.3.2.2 Reconstruction with the Late Approach of Restoration and Improvement

From the 1980s onward, culture-based regeneration has affected regeneration projects all over Europe in a vast wave. Culture-based regeneration and cultural capital 27 have established themselves as successful options in restoration, urban improvement, and renewal (Kearns & Philo, 1993). The most critical aspect of this trend is paying attention to the role of culture, using historical and cultural values as resources for development, and paying attention to people's leisure time.

Ever since policymakers and city officials turned to "culture" as a tool for urban regeneration, the importance of the quality of historical environments as a part of city regeneration innovations and initiatives has become increasingly apparent. Cultural policies were considered to provide economic benefits by creating businesses related to the operators of those businesses, advertising, and promotion, as well as related venues such as cafes, restaurants, etc. Thus, cultural interventions are more than the secondary product of economic growth; they are its driving force (Izadi, 2007).

The museum-building movement in Europe, particularly following the collapse of the Berlin Wall, has emerged as a prominent force in revitalizing collective memories and forging a fresh and distinctive urban identity for cities and their inhabitants. It is a testament to the utilization of cultural instruments in urban restoration and regeneration (Bianchini & Parkinson, 1993). Indeed, culture and cultural initiatives have progressively become significant in urban renewal. Consequently, regeneration can be bolstered by implementing credible and esteemed cultural or artistic projects offering numerous economic and social advantages (Caves, 2005).

In a report by Evans and Shaw in 2004 for the British Ministry of Culture, Media and Sports to design a policy document on the participation of cultural activities in recreation, the types of cultural involvement in different entertainment fields are stated (Table 3.1).

3.3.3 Step Three: A Landscape Approach

Although the recent approach of integrated conservation and development has effectively addressed the limitations of separate conservation and development efforts, it has not yet provided a comprehensive solution to the challenges arising from the interaction between conservation and development. Some experts argue that no overarching urban or regional conservation theory can serve as the intellectual foundation for formulating guidelines on transforming and altering historical sites. Instead, practical endeavors must be reconsidered and incorporated within a flexible and beneficial framework (Bandarin et al., 2012).

From this perspective, the city can be viewed as a dynamic phenomenon that embraces or rejects changes through the lens of the cultural identity of its inhabitants. However, this process unfolds over time, posing potential pitfalls for the integrated conservation and development approach. In other words, the city can be regarded as a self-regulating system that refines itself by subjecting changes to a self-reinforcing filter and monitoring the system. Tuan referred to this refining process within the system as "cultural self-improvement" (Tuan, 1977). This self-improvement acts as a crucial filter, ensuring stability under ideal conditions where public awareness has reached a stage that allows for a correct understanding of changes. This level of awareness surpasses contemporary participatory approaches to conservation and integrated development (UNESCO, 2011). Therefore, considering the landscape's characteristics, the proposed approach is holistic, referred to as

Table 3.1 The theories of city regeneration

Theories related to the sustainable regeneration of historical cities				
No	Theorist	Reference document	Theory	Proposed solution
1	Kevin lynch	*The Image of the City* (1960)	Image of the city	Lynch explored the mental maps people create to navigate and perceive cities, emphasizing the importance of landmarks, paths, edges, districts, and nodes
2	Camillo Sitte	*City Planning According to Artistic Principles* (1889)	Theory: City planning according to artistic principles	Sitte criticized the standardization of urban design and advocated for considering aesthetic and historical aspects, emphasizing the importance of public spaces, squares, and urban forms
3	Lewis Mumford	*The City in History: Its Origins, Its Transformations, and Its Prospects* (1961)	Mumford's work examined the historical development of cities	Highlighting the influence of technological advancements, social dynamics, and urban planning on the evolution of urban environments
4	Jane Jacobs	*The Death and Life of Great American Cities* (1961)	The death and life of great American cities	Jacobs focused on cities' vitality, emphasizing the importance of mixed-use neighborhoods, walkability, and community participation in urban planning decisions
5	Christopher alexander	*A Pattern Language: Towns, Buildings, Construction* (1977)	A pattern language	Alexander proposed a pattern-based approach to urban design, advocating for design elements that enhance human Well-being and create cohesive, organic urban environments
6	Aldo Rossi	*The Architecture of the City* (1966)	The architecture of the city	Rossi explored the relationship between architecture and the city, emphasizing the importance of collective memory, urban artifacts, and the preservation of urban identity
7	Gordon Cullen	*The Concise Townscape* (1961)	The concise townscape	Cullen examined urban design's visual and perceptual aspects, highlighting the significance of townscape composition, focal points, and visual sequences in creating memorable urban environments

(continued)

Table 3.1 (continued)

Theories related to the sustainable regeneration of historical cities				
No	Theorist	Reference document	Theory	Proposed solution
8	Ian L. McHarg	*Design with Nature* (1969)	Design with nature	McHarg advocated for ecological planning and integrating natural systems into urban design, emphasizing the importance of understanding the landscape's ecological processes and incorporating them into sustainable city planning
9	Anne Whiston Spirn	*The Granite Garden: Urban Nature and Human Design* (1984)	The granite garden: Urban nature and human design	Spirn explored the relationship between nature and urban design, emphasizing the role of green spaces, water, and natural elements in creating sustainable and livable cities
10	Carl Steinitz	"Landscape planning: A unified approach" (1990)	Landscape planning: a unified approach	Steinitz developed an approach to landscape planning that considers the interrelationships between natural, cultural, and built environments, emphasizing the need for integrated and holistic strategies in sustaining historic cities
11	James corner	*Recovering Landscape: Essays in Contemporary Landscape Architecture* (1999)	Landscape urbanism	Corner's theory promotes the idea of landscape as a medium for urban transformation, emphasizing the integration of ecological systems, public space, and infrastructure in sustainable urban design
12	Charles Waldheim	*Landscape as Urbanism: A General Theory* (2016)	Landscape as urbanism	Waldheim's theory focuses on repositioning landscape within contemporary urbanism, highlighting the potential of landscape thinking to address ecological, social, and economic challenges in historic cities
13	Kongjian Yu	"Sponge City: Water management and landscape urbanism in China" (2016)	Sponge city	Yu advocates for sponge cities that absorb and utilize rainwater to mitigate flooding, improve water resource management in urban areas, and promote ecological sustainability

(continued)

Table 3.1 (continued)

Theories related to the sustainable regeneration of historical cities				
No	Theorist	Reference document	Theory	Proposed solution
14	Jukka Jokilehto	*A History of Architectural Conservation* (1999)	An integrated and holistic approach to preserving historic cities	He presents the theory of historically valuable structures as "monuments" and discusses the restoration movements and theories that emerged during different periods, including the renaissance era and the seventeenth-century civil war in Europe
15	Brown (2018) and Wilson (2021)	"A spatial overview of the global importance of indigenous lands for conservation" (2018)	Development of innovative approaches to preserving historic landscapes	Highlight the development of innovative approaches to preserving historic landscapes that incorporate environmental conservation, cultural diversity preservation, and community participation
16	Johnson (2007) and Smith (2015)	*The Archaeology of Urban Landscapes* (2014)	Preserving historical landscapes	Discuss the development of laws and regulations for preserving historical landscapes, particularly after world war II

Source: Author

the landscape approach. It perceives conservation and development not as two separate fields or interdisciplinary categories but as a broad domain in which development aligns with the path of conservation or, in other words, where development becomes an integral part of the self-improvement process of protection, which is influenced by development. The flexibility of this approach is contingent upon development (Nocca et al., 2018).

By prioritizing the preservation of processes rather than the products themselves, the aim is to foster development that undergoes the filter of cultural self-purification, which is a product of a comprehensive understanding of the ecosystem rather than fragmented perspectives. Through this process, opportunities for creativity can emerge, benefiting the current generation, while safeguarding the rights of future generations. In this idealistic approach, conservation seeks not only to preserve the existing historical city but also to enable its sustainable use within the natural environment, transcending the chronological boundaries of the town and considering its entire ecological foundation. By analyzing various aspects of the historic urban landscape, one can uncover the natural-cultural processes that shaped and sustained it, thus ensuring the continuity of these processes and the possibility of new outcomes (Eshrati & Doulabi, 2021).

This approach views the historical urban landscape as a generative phenomenon rather than a static artifact in a city museum. The filter of self-purification, which restricts interference with heritage, acts as a guarantee for preserving the legacy for future generations. It should be noted that this filter possesses unique characteristics within each society and is effective only for the culture and microclimate it serves. Therefore, a universally fixed filter cannot be achieved, and cultural filters are as diverse as indigenous communities. This aligns with the recent conservation organizations' approach of seeking local solutions rather than global ones. It is worth mentioning that realizing ideal conditions, where public awareness has reached a stage of understanding the type and limits of changes, is considered an "ideal" scenario. However, in the bipolar world we inhabit, these ideal conditions are not entirely attainable; thus, the system presented in this research may not be fully realized. Increasing public awareness is necessary for effective management strategies, providing a foundation for the sustainability of both protection and development. However, recognizing the intricate interconnectedness of the historic urban landscape and considering each component within its management process is essential for sustaining the overall process.

3.4 Summary and Conclusion

The view of landscapers as a new approach to reviving and recreating historical contexts as a place can help to protect the tangible and intangible values of these places and the continuity of identity and, on a larger scale, preserve the narrative of the city's history. This can effectively increase citizens' sense of belonging to the town. The purpose of belonging is an influential factor in the continuity and stability of a place. The landscape of historical sites is a rich text, and any intervention, change, and interaction with it are impossible without reading and understanding the relevant narrative. Applying the landscape approach in recreating and revitalizing historical contexts can help to continue the possibility of the correct reading of the landscape text and perception of the historical narrative of the place in historical contexts and strengthen the sustainability of historical cities (Khorasgani et al. 2023).

The landscape approach in regenerating historic cities is a comprehensive and holistic strategy recognizing the inherent interconnections between natural and built environments and urban spaces' social and cultural dimensions. This approach emphasizes the integration of green infrastructure, public spaces, and heritage conservation in urban planning and design, aiming to enhance the quality of life, sustainability, and resilience of historic cities. It involves the careful analysis and understanding of historical context, urban morphology, and sociocultural values, as well as the involvement of local communities and stakeholders in the decision-making process. Scholars and practitioners have highlighted the significance of the landscape approach in regenerating historic cities, as it fosters a harmonious relationship between heritage preservation, environmental sustainability, and urban development (Benedict & McMahon, 2006; European Landscape Convention, 2000; UNESCO, 2011).

Landscape architecture is a burgeoning discipline that explores the environment's visual and spatial aspects, effectively shaping individuals' sense of belonging (Fjeld, 2018). The sense of belonging plays a pivotal role in the continuity and vitality of a place (Low & Altman, 1992). Historical sites, with their rich narratives, can be interpreted through the lens of landscape (Corner, 1999). By offering a fresh perspective, landscape architecture profoundly influences the sense of belonging (Francis, 2003). It deciphers and elucidates the symbols and phenomena present in the living environment, thereby managing and recreating the sensory-perceptual elements that contribute to the texture of a place (Kaplan & Kaplan, 1982). History serves as a guiding force (Tilden, 1977).

Landscape architecture is an intermediary between humans and their surroundings, acting as both an indicator and catalyst of semantic components within historical contexts (Cosgrove, 2004). Through its narrative approach, landscape architecture presents a distinctive expression of a site, enhancing the sense of place and fostering a stronger sense of belonging (Corner, 1999). It establishes a connection between individuals and their environment over time, facilitating the development of emotions, perceptions, and collective memories that encourage social engagement (Nasar, 1998). This interpretation becomes intertwined with the identity of a place, reinforcing the sense of belonging experienced by individuals (Relph, 1976).

The landscape approach is a comprehensive and holistic perspective that considers various factors and influences within historical contexts, aiming to achieve a comprehensive continuity and recreation of historical thoughts and insights (Smith, 2008). By examining landscapes from a human-centered viewpoint, it considers factors such as a sense of belonging, personal and collective memories, a sense of participation, a sense of place and time, and an increase in sensory richness, all of which contribute to social vitality (European Landscape Convention, 2000).

In historical landscapes, each component plays a crucial role in maintaining dynamism and stability, ensuring the continuity of these biological collections. By managing these factors effectively, the landscape approach facilitates preserving and enhancing historical landscapes for future generations (Jokilehto, 1990).

Furthermore, similar cases can be found where the landscape approach has been successfully applied. For instance, the preservation and revitalization of heritage sites such as the Historic District of Old Québec in Canada have adopted the landscape approach to protect the authenticity and integrity of the historical environment (Gumart, 2018). By considering the historical context and its associated elements, including architecture, urban layout, and cultural values, the landscape approach ensures a comprehensive and holistic management strategy for heritage conservation (Fig. 3.1).

Overall, the landscape approach emphasizes the importance of considering all relevant factors and influences within historical contexts to achieve comprehensive continuity and recreation of historical thoughts and insights. By recognizing the significance of factors such as a sense of belonging, memories, participation, place and time, and sensory richness, this approach promotes social vitality and enhances the management and preservation of historical landscapes (Smith, 2014; European Landscape Convention, 2000).

In conclusion, the regeneration of historical cities has undergone significant evolution, with different approaches and methods used to revitalize these cities and

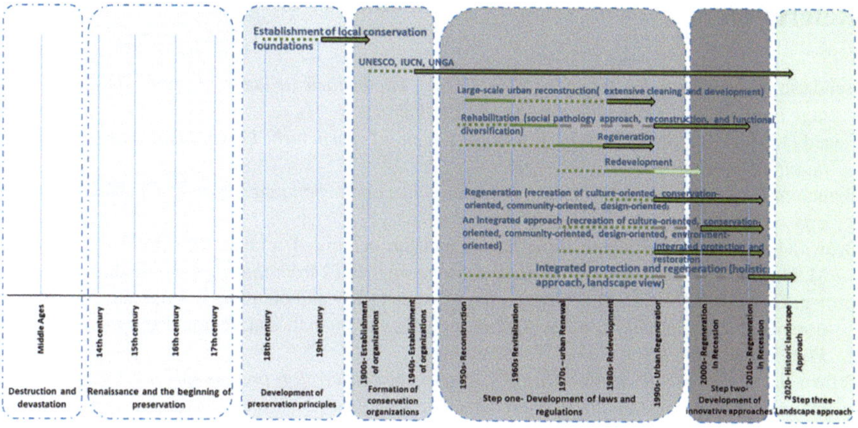

Fig. 3.1 The evolution of the historical city's regeneration. (Source: Author)

preserve their cultural heritage. The historical focus on physical restoration has shifted to consider social and economic needs, and the importance of community involvement and participatory planning is now recognized. Historic cities hold immense value as a part of our cultural heritage, offering glimpses into our past and fostering a sense of continuity between generations. However, many of these cities have suffered from urban decay and neglect. To counteract this trend, regeneration efforts have been launched to revitalize these cities and create vibrant, sustainable communities. The evolution of these efforts, from the mid-twentieth century to the present day, has witnessed significant shifts in approach.

The evolution of historical city regeneration highlights the necessity of adopting a holistic and comprehensive method that encompasses all aspects and needs of sustainability. Examining the theories over time reveals that contemporary periods that have embraced holistic approaches have shown tremendous success, leading to more recommendations by theorists and relevant organizations in these fields.

The landscape has emerged as a holistic approach to providing solutions for the sustainable regeneration of historical cities. By embracing the landscape approach, historical cities can achieve sustainability and continuity.

The landscape approach remains a guiding principle for many cities in regenerating historic areas. While new technologies and approaches continue to emerge, the landscape approach has firmly established itself in creating sustainable and vibrant communities while preserving our cultural heritage. By prioritizing the community's needs, the landscape approach ensures that historic cities remain relevant and valuable for future generations. The evolution of historical city regeneration will persist, but the principles of the landscape approach will continue to serve as a guiding force in these efforts.

References

Bandarin, F., Van Oers, R., & Licciardi, G. (2012). *The historic urban landscape: Managing heritage in an urban century* (pp. 68–70). Wiley.

Benedict, M. A., & McMahon, E. T. (2006). *Green infrastructure: Linking landscapes and communities*. Island Press.

Bianchini, F., & Parkinson, M. (1993). *Cultural policy and urban regeneration: The West European experience*. Manchester University Press.

Brown, J. (1999). The English Renaissance and the origins of archaeology. In P. P. A. Funari, M. Hall, & S. Jones (Eds.), *Historical archaeology: Back from the edge*. Routledge.

Brown, C., & Dühr, S. (2018). Understanding sustainability and planning in England: An exploration of the sustainability content of planning policy at the national, regional and local levels. In *Planning in the UK* (pp. 257–278). Routledge.

Brown, L., Desai, A., & Russell, S. (2010). *Urban regeneration and social sustainability: Best practice from European cities*. Wiley-Blackwell.

Brown, L., O'Dwyer, B., & Lyons, M. (2018). Measuring the socio-economic impacts of urban regeneration on deprived communities. *Urban Studies, 55*(14), 3114–3132.

Bruno, A. (2018). World heritage and sustainable development: The role of local communities in deciding the future of the Bamyan Valley (Afghanistan). In P. B. Larsen & W. Logan (Eds.), *World Heritage and sustainable development: New directions in world heritage management*. Routledge.

Caves, R. E. (2005). *Creative industries: Contracts between art and commerce*. Harvard University Press.

Corner, J. (1999). *Recovering landscape: Essays in contemporary landscape architecture*. Princeton Architectural Press.

Cosgrove, D. (2004). Landscape and the mythologies of identity. In M. Bender & G. W. Huggan (Eds.), *Landscape and identity: Critical essays*. Peter Lang.

Delafons, J. (1997). Sustainable conservation. *Built Environment (1978-), 23*, 111–120.

Eshrati, P., & Doulabi, P. (2021). Analysis of the historic urban landscape of Isfahan based on the cultural and natural foundations from the beginning to the end of the Safavid era. *Armanshahr Architecture & Urban Development, 13*(33), 153–166.

European Landscape Convention. (2000). *Council of Europe*. Council of Europe Publishing.

Fjeld, H. (2018). Landscape architecture as a discipline. In D. P. Miller & J. Tinus (Eds.), *Landscape theory in design*. Routledge.

Francis, M. (2003). Place and placelessness in landscape design. In M. Francis (Ed.), *The meaning of gardens*. The MIT Press.

Gumart, C. (2018). Quebec: A History of Culture and Politics.

Hajipour, K. (2011). Urban renewal and regeneration policies: A case of Tehran, Iran. *Journal of Geography and Regional Planning, 4*(9), 515–524.

Hajipour, K. (2018). Urban regeneration and the emergence of creative cities. In H. Zhang, J. Zhou, & J. Li (Eds.), *Creative economy and culture*. Springer.

Hall, P. (1998). *Cities in civilization: Culture innovation, and urban order*. Weidenfeld & Nicolson.

ICOMOS, I., & Mission, I. C. O. M. O. S. (1994). About Icomos. NATIONAL GEOGRAPHIC, 1983(1993).

Izadi, M. (2007). Preservation of historic urban landscape: A new approach in urban planning. *Journal of Social Sciences, 3*(4), 194–199.

Johnson, B. (2007). Systems of innovation, the urban order and sustainable development. *Waste Management & Research, 25*(3), 208–213.

Johnson, M. H. (2012). *Archaeological theory: An introduction*. Wiley.

Jokilehto, A. (1990). The preservation of historical landscapes: From destruction to the establishment of preservation laws in. *Journal of Cultural Heritage, 1*(1), 5–14.

Kalantri, A. P. (2009). *Urban regeneration: Policies and strategies*. New Age International.

Kaplan, R., & Kaplan, S. (1982). *Cognition and environment: Functioning in an uncertain world.* Praeger.
Kearns, G., & Philo, C. (1993). *Selling places: The city as cultural capital past and present.* Pergamon Press.
Khorasgani, A. M., Villalobos, M. H., & Eskandar, G. A. (2023). Sustaining historic cities: An approach using the ideas of landscape and place. *ISVS e-Journal, 10*(1), 320–332.
Larkham, P. J. (1999). Tensions in managing the suburbs: Conservation versus change. *Area, 31*(4), 359–371.
Low, S. M., & Altman, I. (1992). Place attachment: A conceptual inquiry. In I. Altman & S. M. Low (Eds.), *Place attachment.* Plenum Press.
McDonald, G. (2009). *Regeneration and neoliberalism.* Routledge.
Miller, N. F., & Gleason, K. L. (Eds.). (1995). *The archaeology of garden and field.* University of Pennsylvania Press.
Nasar, J. L. (1998). *The evaluative image of the city.* Sage.
Nocca, F., & Girard, L. F. (2018). Towards an integrated evaluation approach for cultural urban landscape conservation/regeneration. *Region, 5*(1), 33–51.
Pendlebury, J., & Miciukiewicz, K. (2015). Metropolitan regions, historic centres and cultural heritage: The case of the United Kingdom. *CyTET–Ciudad y Territorio Estudios Teritoriales, 17*(184), 219–236.
Pourahmad, A., Habibi, K., & Keshavarz, M. (2011). The evolution of urban regeneration as a new approach in worn-out urban contexts. *Tehran: Iranian, in Islamic City Studies, 2*(5), 5–18.
Relph, E. (1976). *Place and placelessness* (Vol. 67). Pion.
Roberts, P. (2000). The evolution, definition and purpose of urban regeneration. *Urban regeneration: A handbook, 1*, 9–36.
Roberts, P. (2008). The evolution, definition, and purpose of urban regeneration. In P. Roberts & H. Sykes (Eds.), *Urban regeneration: A handbook.* SAGE Ltd.
Sahizadeh, F., & Yazdi, M. M. (2010). The role of conservation and development in urban planning. *International Journal of Environmental Science and Development, 1*(5), 409–413.
Smith, A. D. (2008). The shifting landscapes of nationalism. *Studies in Ethnicity and Nationalism., 10*(5), 51–77.
Smith, M. L. (2014). The archaeology of urban landscapes. *Annual Review of Anthropology, 43*(1), 307–323.
Smith, R. M. (2015). Planning for urban sustainability: The geography of LEED®–Neighborhood Development™(LEED®–ND™) projects in the United States. *International Journal of Urban Sustainable Development, 7*(1), 15–32.
The English Heritage Visitors' Handbook 1998–99. United Kingdom, English Heritage, 1998.
Thompson, M. W. (2010). *Ruins: Their preservation and display: A Colonnade book.* British Museum Press.
Tilden, F. (1977). *Interpreting our heritage.* Chapel Hill: The University of North Carolina Press.
Tuan, Y. F. (1977). *Space and place: The perspective of experience.* University of Minnesota Press.
UNESCO. (1972). *Convention concerning the protection of the world cultural and natural heritage.* Retrieved from https://whc.unesco.org/en/conventiontext. UNESCO.
UNESCO. (2011). *Historic urban landscape: Recommendation on the historic urban landscape, UNESCO World Heritage Centre.* Paris: UNESCO World Heritage Centre.
Willson, M. A. (2006). *Technically together: Re-thinking community within techno-society* (Vol. 28). Peter Lang.
Wilson, B. D. (2021). *Futureproof city: Ten immediate paths to urban resilience.* Routledge.
Wilson, J., & Groth, P. (Eds.). (2006). *Everyday America: Cultural landscape studies after J.B. Jackson.* University of California Press.

Chapter 4
Landscape Approach for Historic City Sustainability

4.1 Introduction to the Landscape Approach

The landscape concept comprises three interconnected and unified elements: site, viewpoint, and image. In the idea of landscape as a perceptual phenomenon, the landscape is defined as a perceived space from the human point of view (Collot, 2011). The landscape is another form of place (Mansouri, 2008) and the product of interaction between humans and the external environment (Berque, 2010). The landscape has the nature of place and the product of human experience in space, a dynamic and relative physical-mental phenomenon that takes shape in interaction with history and character (Mansouri, 2005). Landscape, as a new branch of sciences that analyzes and studies the human living environment, provides a non-Cartesian interpretation of the world, a systemic perspective, and a redefined understanding of "intrinsic material content" through a reconsideration of human thinking methods (Mansouri & Farzin, 2016). The essential principle in the landscape is forming or integrating form and meaning. In other words, its holistic view is the most significant point that distinguishes the landscape approach from other sciences and orientations and defines a clear difference. The landscape approach, like the concept of landscape, can be expressed in various ways by emphasizing each of its components: human, environment, and their relationship; meanwhile, common grounds can be established between different approaches.

Landscape, as an intermediary and interface between humans and places, Fig. 4.1, is a concept that has been addressed in many fields related to the environment, including architecture, urban planning, landscape archaeology, environmental psychology, and sociology. Landscape can be defined as the result of the integration of human activities and nature, representing the evolution of human societies over time. As an intermediary and interface between humans and places, the landscape reflects perspectives, values, cultures, and human experiences in spatial arrangements (Heydari, 2020). As a phenomenon in Fig. 4.2, the landscape

A. Moazzoni Khorasgani, *Using Data Science and Landscape Approach to Sustain Historic Cities*, https://doi.org/10.1007/978-3-031-68161-5_4

Fig. 4.1 Interaction of landscape with humans and place. (Source: Author)

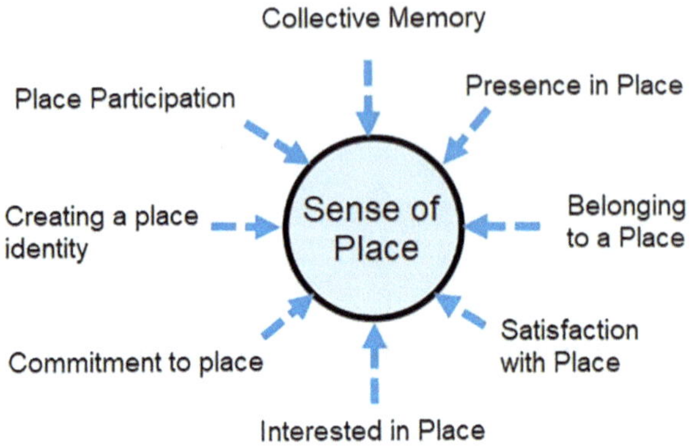

Fig. 4.2 The sense of place and factors related to it. (Source: Authors)

takes shape in the exchange between the tangible and perceptible characteristics of the environment on one side and patterns, cultural symbols, and the observer's mental capabilities on the other (Ansari, 2017).

Therefore, the role of landscape as an intermediary and interface between humans and places lies in facilitating the interaction and exchange of information between a place's tangible and mental layers. This communicative role resulting from interaction can generate a sense of attachment to a place, demonstrating the highest level of human-place relationship.

4.2 Principles of Landscape Approach

4.2.1 Definition of Place

The place is a spatial entity that carries meaning for individuals or groups. People react to their environment based on the perceptions it creates for them (Rapoport, 1982). The place encompasses various characteristics that together make diverse potential meanings. As a collection, the place has the potential to facilitate a process in which individuals develop and maintain a sense of place, belonging, and identity. Place resides at the center of the world or social realm. Places take shape or are constructed when human interactions with their physical environment occur. Places are defined by physical entities and social and psychological activities and processes within them (Stedman, 2002). The place is not merely comprised of physical locations; instead, it is filled with symbolic meanings, emotional attachments, and sentiments that individuals have about a specific set of surroundings (Dominy, 2002).

The physical components, activities, and meanings are the constitutive elements of places, forming both a place's tangible and mental layers. While the first two elements may be relatively easy to comprehend, understanding the meaning component is much more challenging. The meanings of places may have roots in the physical arrangement of objects and activities, but meanings are not solely derived from these factors; they are heavily influenced by people's initial expectations, experiences, intentions, and purposes as observers and users of places (Relph, 1976). Thus, the environment represents the shared mental constructs of individuals (Carmona, 2021).

Lynch defines the identity of a place as the distinguishing factor that determines it from other places, making it recognizable as a unique entity (Lynch, 1972). Therefore, the identity of a place can be adequately understood through individualization. The identity of a place focuses more on the mental and emotional satisfactions of human beings rather than their bodily needs. Both individual and collective values shape the identity of a place and evolve, gaining depth, expansion, and change (Relph, 1976). Schultz views the study of place as the study of events and incidents that occur within it, and he believes that a collection of events, forms, colors, and textures shapes the character of a place or its identity. He establishes a connection between architecture, place, and cultural identity and argues that the experience of place is an experience of the meaning of place (Norberg-Schulz, 1981). Landscape can serve as a narrator and mediator of the human experience of place and, therefore, has a direct relationship with the sense of place and its continuity. Table 4.1 examines various opinions of different experts regarding the components and influential indicators of the sense of place, which validate the relationship between landscape and the sense of place.

Sense of place is an emotional bond that develops through an individual's attachment, awareness, belonging, satisfaction, and commitment to a place (Shamai, 1991; Kaltenborn, 1998). The perception of the sense of place varies over time. It depends on individuals' understanding, their level of presence and engagement in the place, and the perceived quality and clarity of the landscape (Kashi & Bonyadi, 2013). Understanding the influential factors at different levels of the sense of place is crucial to enhancing the sense of place and ultimately creating a sense of place identity. These factors can be considered as criteria for measuring the existence of the sense of place and, finally, place identity. As discussed, at the initial levels of the sense of place, the focus is primarily on physical factors and physical presence in a place. In the intermediate levels of the sense of place, participation and attachment to the place through activities play a significant role. At the final levels of the sense of place, a sense of unity and complete commitment to the place can be observed due to the meanings shaped in the minds of individuals—Fig. 4.2.

The sense of place describes an individual's cognitive, emotional, and functional relationship with a specific location. It often leads individuals to develop a sense of belonging or to perceive meanings associated with that place (Tuan, 1993). Therefore, the sense of place is closely related to the landscape, and the mental layers of the landscape form the influential factors of the sense of place within a location.

Table 4.1 Definitions of place

No.	Researcher	Practical components strengthen the sense of place
1	Rapaport	Physical signs consider the organization of the environment dependent on the organization of the four elements of space, time, communication, and meaning. They generally consider it equivalent to a series of communications and consider the relationship between the environment and people meaningful. The individual's attitude and the discovery of environment connections—cultural symbols and symbols, social experiences—emphasize the role of sociocultural components in forming the physical environment; with the theory of nonverbal communication, the environment is a collection of fixed, semifixed, and mobile physical elements. It means a person infers his meaning and concept from the environment and its elements based on his cultural codes and ideas. From his point of view, in this approach, the place becomes a symbolic place where each element symbolically shows a part of the social culture of its environment
2	David Simon	A place is an environment that collects human experiences, actions, and meanings both spatially and temporally. Among the components of meaning, there is a common sign of the three principles of differentiation, continuity, and communication, which are surrounded by a circle. The location model is shown
3	Kanter	According to Kanter's model, the place consists of form, activity, and meaning and has three dimensions: physical, functional, and perceptual-cognitive. The physical characteristics of the place's environment are the most tangible and are the most critical components of the place compared to the other two elements (activity and imagination)
4	Relph	He considers the place to consist of natural and artificial environments, human actions, and meanings that can be experienced on different scales
5	Norberg-Schulz	He considers the spirit of the place to be the essence of communication between man and place. The sense of place is found in places with a specific and distinct character, and the environmental character is made of tangible things with materials, shape, fabric, and color. It creates unity, and the space must have a perceptible identity and be recognizable, memorable, and visible to create a sense of place. This kind of sense of place also brings a sense of belonging
6	Kormuna	He considers the place to be the result of three factors: form, activity, and individual imagination, which lead to its formation in an interactive relationship. In addition, they have stated that parts of the city remain unchanged over time in terms of the continuity of the signs of the place. They create a sense of continuity in the place
7	Jennifer Kross	He considers the factors affecting the sense of place to be the relationship with the place and the sense of community. He categorizes the connection with the place into autobiographical, spiritual, ideological, narrative, material, and forced dependence forms related to factors. Identity, interiority, and satisfaction create five levels of sense of place
8	Altman	Cultural beliefs related to the environment, the scale of the climate, exclusivity, and accessibility are influential factors in strengthening the sense of place

(continued)

Table 4.1 (continued)

No.	Researcher	Practical components strengthen the sense of place
9	Fallahat, Kamali, and Shahidi (2017)	Sense of place means people's subjective perception of the environment and the factor of transforming a space into a place with unique sensory and behavioral characteristics for certain people
10	Kashi and Baniadi (2013)	Participation and attachment to the place through its activities lead to a stronger relationship with it. At the last level of the sense of place, the feeling of being one and fully committed to the place is observed due to the meanings formed in people's minds.
11	Prozhanski	He considers the factors influential in strengthening the sense of place to be an attachment to the place—the identity of the place, interaction with the environment, a close connection between the person and the environment
12	Kersul	Place is a combination of matter, meaning, and action. In his drawing model, place connects a person with activities and functions that are geographically located. This phenomenon starts at the personal level and expands to the level related to lived experiences

Source: Author

The relationship between the sense of place, landscape, and humans is a complex and multidimensional subject studied in various fields such as geography, psychology, art, and architecture. The sense of place refers to humans' cognitive and emotional perception of a space or environment, which creates attachment, identity, and meaning for individuals or communities. The landscape plays a significant role in shaping the sense of place. It presents a beautiful image of space. It stimulates the five senses, providing sensory and spiritual experiences for humans. Through their interaction with space and the landscape, humans also reflect their cultural, historical, social, and personal factors, producing the spirit of a place. Therefore, the relationship between the sense of place, landscape, and humans is a two-way relationship that depends on the environment's physical and nonphysical aspects and humans' mental and emotional factors (Kolvir et al., 2019). One of the influential factors in the sense of place is collective memory, which will be discussed further.

4.2.2 Mindscape

In an increasingly interconnected world, sustainability has become paramount, touching upon diverse facets of human existence, from environmental conservation to socioeconomic stability. At the heart of this multifaceted challenge lies the human mindscape—the intricate interplay of beliefs, values, perceptions, and cognitive processes that shape our decisions and actions. This research paper's central objective is to understand the mindscape's profound influence on sustainability. To unravel this intricate relationship, we delve into a rich tapestry of theoretical frameworks drawn from psychology, sociology, and environmental studies. By exploring how our mental landscapes impact our collective pursuit of sustainability, we aim to shed light on the critical role that cognitive and perceptual dimensions play in shaping our global future.

4.2.2.1 Factors Affecting Mindscape

4.2.2.1.1 Collective Memory

Longman's dictionary defines memory as "A person's ability to remember things, places, experiences" (Yorkey, 1979), while Webster's dictionary defines it as "the ability or process of recalling or reproducing what has been learned or memorized, especially using associative mechanisms of meaning." Based on these definitions, memory can be defined as the recollection of past events and experiences by a person, including people, objects, and places, and the use of associative mechanisms to attach meaning to these recollections.

Memory can be individual or collective, shaped by events recognized and shared by a community. The more individuals remember and share these memories, the greater their significance and impact. Maurice Halbwachs was the first to propose the concept of collective memory, which Aldo Rossi, the Italian neo-rationalist architect, later introduced to architecture and urban planning literature. Rossi saw a city's history as the soul of the city and the manifestation of its collective memory. In other words, a city's identity is defined by its memory, which is connected to objects and places. The relationship between the city and its citizens creates the dominant image of the town, with certain artifacts becoming a part of the collective memory and shaping its identity. Collective memory plays a crucial role in shaping the mindscape of a city. It involves shared narratives, histories, and cultural heritage contributing to the resident's sense of identity and connection to their surroundings. Preserving and promoting collective memory through heritage preservation, commemorative practices, and storytelling can strengthen social ties, encourage community engagement, and enhance the sustainability of urban contexts (Huyssen, 2003; Assmann, 2011).

Memories need a time reference to form. The passage of time and the course of historical development are essential factors in remembering events. Also, collective memories, like the historical memory of a nation, are passed from one generation to another due to the history and time of living in a city, strengthening collective memories (Manzo & Devine-Wright, 2013).

Memories need a spatial reference to form. With their physical and functional structure and spatial connections, urban places and spaces are considered spatial references for creating memories in urban areas. According to Aldo Rossi, buildings and elements are formed as reference points in residents' minds. According to Halbwachs, any change in the physical appearance of urban spaces causes a severe change in the mental image of city residents and affects their sense of collective memory. As long as the location of buildings, stones, and trees in a city has not changed, the residents' mental image of that space has not changed, and there is a sense of belonging to a place and the collective memory of its residents (Ward et al., 2016).

The culture of social interactions and collective memories takes place in a context that is the place of social interactions. For the formation of memories, in addition to the physical background of the city, there must also be a social-cultural background so that people with diverse connections can form collective interactions, which leads to the formation of collective memories. A social relationship may occur between a group of people. Still, a collective memory is not formed, but a collective memory cannot happen without a social relationship. With the manifestation of these communities, public arenas come to life, relationships become more profound, and the passage of time gives meaning to collective memories (Casakin & Bernardo, 2012).

Mental and individual experiences, including job, education, and the degree of connection of a person with the environment, as well as the way he or she perceives the environment, are among the other components that form and influence collective memory because these two factors play an essential role in the formation of a person's mental image toward the environment. In other words, the quality of sight and scenery is how the audience interacts cognitively in urban environments.

According to MirMoqtadaei (2004), collective memories create a sense of belonging to a place, social cohesion, participation, and a sense of place. Additionally, the revitalization project includes economic, social, environmental, cultural, physical-spatial, functional, and managerial dimensions, and collective memory can play a role in each of these aspects of urban revitalization. Regarding the economic dimension, a sense of belonging created through collective memory can encourage investment among investors with a special connection to the site. In the social dimension, collective memory can strengthen the spirit of participation, which can increase the desire to invest and participate in the renewal of the urban economy, improve the physical quality of the environment, and facilitate cooperation with the public sector. Moreover, collective memories can enhance the protection and preservation of the environment and create social cohesion, strengthening social life and increasing social security. Collective memories can also create and promote a sense of place, attracting tourists and developing tourism. By creating impressive spaces that create a coherent and memorable image in mind, collective memories play an essential role in enhancing the sense of place. Hence, collective memories have various effects on revitalizing urban tissues and biological collections (MirMoqtadaei, 2004).

4.2.2.1.2 Belonging Sense

Belonging sense to a place means that individuals identify with where they were born and raised. This identification leads to an emotional connection with the place, and individuals consider themselves a part of the area, creating a unique role based on their experiences of the place's signs, meanings, and functions (Ryan, 2006). This emotional connection, known as spatial attachment, is the foundation of positive emotional interactions and communications with the environment (Ryan, 2006) and is influenced by individual, collective, cultural, and social characteristics (Ryan, 2006). Spatial belonging can be understood in several dimensions.

The first dimension is spatial identity, which refers to the personal relationship between individuals and their environment and how it contributes to their identity. Proshansky's concept of spatial identity suggests that individuals define their identity in their physical environment by distinguishing themselves from others and preserving their sense of identity with people in the area (Manzo & Devine-Wright, 2013). Place identity is one aspect of spatial identity that emerges and develops in the physical environment through a pattern of opinions, preferences, feelings, values, and goals (Parkins & Craig, 2006).

The second dimension is spatial dependence, which is related to the potential of a specific place to satisfy an individual's needs and goals compared to other sites. Spatial dependence is about individuals choosing a place that meets most of their expectations and needs.

The third dimension is an emotional connection, which is dynamic and expressive of a particular environment and context. Emotional attachment is associated with passionate responses to the context and environmental conditions. People are not just connected to the physical environment but also feel a sense of belonging to the people who live there. A place's social and cultural relationships are crucial for developing social capital and promoting long-term residence. Studies show that social bonds with friends and relatives are one of the most important and stable sources of spatial attachment. The sustainability of urban contexts goes beyond social and cultural aspects and extends to environmental considerations. The mindscape of a city should also encompass a commitment to ecological balance, resource efficiency, and resilience in the face of climate change and other environmental challenges. Emphasizing sustainable practices, green infrastructure, and environmental stewardship within the mindscape contributes to cities' long-term viability and livability (Beatley, 2011; Folke et al., 2005).

4.2.2.1.3 Social Participation and Its Relationship with Spatial Belonging and Social Capital of Social Participation

The emotional involvement of individuals in group situations is a crucial element in strengthening societies and empowering communities, motivating individuals to support and guide each other in achieving group goals. The social characteristics of a neighborhood also play an essential role in explaining the cohesion of its residents and their desire to participate in community activities. Citizen participation is critical for developing and improving all aspects of urban life, as it allows city managers to address social problems and shortcomings better, ultimately creating a better environment (Mohammadi, 2010). Citizens' participation in city affairs can have positive results, reducing city management's issues and challenges, particularly for municipalities with limited resources and specialized personnel, leading to resource savings (Cunningham, 1972). Social participation refers to the active involvement of individuals and communities in decision-making processes, governance, and local initiatives. It encompasses efforts to facilitate neighborhood management, attract investment, address environmental challenges, and promote sustainable development. Encouraging and enabling social participation fosters a sense of ownership, empowerment, and collective responsibility, leading to cities' sustainable growth and well-being (Innes & Booher, 2003).

4.2.2.1.4 Sense of Place

The sense of place refers to people's mental perception and conscious feelings toward their environment, leading to a unique sensory and behavioral connection with that place. This concept not only provides a feeling of comfort but is also a cultural concept that is highly valued by society as it reinforces their social and cultural relations within a place, reminding them of their past experiences and aiding in developing their identity (Hesami et al., 2022). Therefore, the sense of place is vital in creating a strong emotional attachment between people and their environment, influencing their behaviors and shaping their experiences within a particular area. Place identity refers to a specific location's unique characteristics, meanings, and symbols. It encompasses the emotional and psychological attachment that individuals have toward their environment. Building a solid sense of place identity and fostering a sense of belonging within a community enhance social cohesion, encourage active participation, and contribute to the long-term sustainability of cities (Proshansky et al., 1983; Relph, 1976).

The influence of collective memory on social participation, sense of place, and sense of belonging can be intricate and mutually reinforcing. Collective memory refers to the shared understanding, narratives, and representations of the past within a community or society (Halbwachs, 1992). It shapes individuals' collective identity and sense of belonging within that community. In turn, social participation, sense of place, and sense of belonging are influenced by collective memory through various mechanisms.

Collective memory plays a crucial role in fostering social participation within a community. Shared memories and historical narratives create a sense of collective identity and solidarity among community members. These memories can generate a shared sense of purpose, common goals, and a collective vision for the future. As a result, individuals are more likely to actively participate in community activities, initiatives, and decision-making processes, contributing to the sustainability of cities (Cohen, 2009).

Collective memory contributes to forming and preserving a strong sense of place. The historical events, stories, and symbols associated with a particular location become embedded in the community's collective memory (Lewicka, 2011). These shared memories and meanings shape the cultural, social, and emotional attachments to the place, enhancing the sense of place. A strong sense of place fosters a sense of identity, attachment, and pride among community members, promoting a desire to protect and sustain their environment.

Collective memory also influences the sense of belonging within a community. Shared memories and historical narratives provide a shared heritage that connects individuals to their community's past (McAdams, 1989). This shared heritage creates a sense of continuity, shared experiences, and a feeling of being part of a larger collective. The sense of belonging derived from collective memory strengthens social cohesion, trust, and social capital within the community, which are essential for sustainable urban development (Bonaiuto et al., 2003).

In summary, collective memory influences social participation, sense of place, and belonging in cities by shaping collective identity, fostering social cohesion, and creating a shared vision for the future. Social participation, in turn, contributes to constructing and preserving collective memory. The sense of place and belonging are reinforced through the shared memories, narratives, and symbols associated with a community's past. This interplay between collective memory and these factors is crucial for promoting cities' sustainability by fostering community engagement, attachment to place, and a sense of collective responsibility. To enhance their relevance and comprehensiveness, the findings from the questionnaires were further examined and analyzed in conjunction with the research conducted in the initial phase. As a result, the following outcomes and implications emerged.

The field of landscape architecture, especially mindscape, provides valuable insights into how the landscape impacts architecture and place, ultimately fostering a sense of belonging among individuals (Smith, 2011). This sense of belonging plays a pivotal role in the continuity of a site, and historical places possess rich narratives that can be interpreted through landscapes (Khorasgani & Haghighatbin, 2023).

The mindscape emerges as a critical element in sustaining cities within this framework. Several factors contribute to safeguarding the mindscape, including collective memory, the sense of place, belonging, and social participation (see Fig. 4.3).

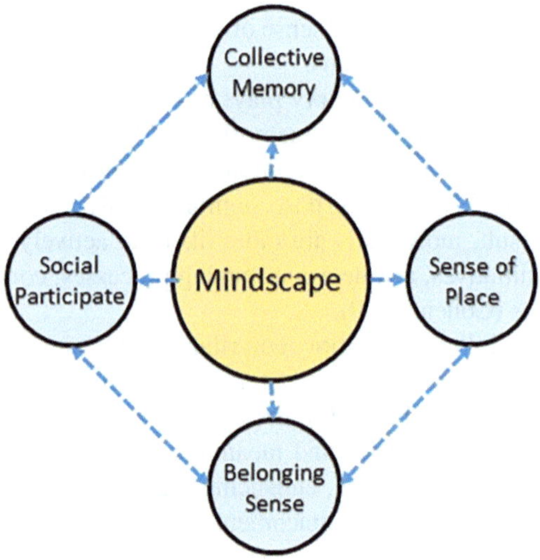

Fig. 4.3 Mindscape and its factors. (Source: Author)

Each of these factors exerts a remarkable influence on the mindscape, which can elucidate the signs and phenomena present in the physical environment, thereby influencing people's sensory-perceptual experiences (Lynch, 1960). The mindscape is an intermediary between individuals and their surroundings, serving as an indicator and catalyst for the semantic components within historical cities (Norberg-Schulz, 1980). Through active engagement, the mindscape presents a distinct expression of a place by narrating the sense of place and belonging (Cresswell, 2014).

The Chicago Bronzeville neighborhood is a significant example of how the mindscape can impact social and physical stability and attract investments. In recent years, the creation of mental images has substantially affected the community, leading to increased acquisitions in this neighborhood (Hou, 2010).

The mindscape concept holds the utmost importance for sustainable cities and communities. It fosters a sense of belonging, collective memories, and factors that drive social participation (Gehl, 2013). Preserving collective memory, the importance of place, belonging, and social involvement are crucial for sustaining the mindscape (Khorasgani et al., 2023). The case study of the Chicago Bronzeville neighborhood vividly demonstrates the mindscape's significant impact on social and physical stability and attracting investments (Carmona, 2010; Florida, 2002). By comprehending and harnessing the concept of mindscape, cities and communities can create sustainable environments that promote social and physical well-being (Gibson, 1977). The interplay between the theoretical frameworks explored in this paper sheds light on how the mindscape influences the sustainability of cities. The discussions surrounding place identity, collective memory, social participation, and environmental sustainability highlight how these factors can be leveraged to promote the development and well-being of urban contexts (Beatley, 2011). By prioritizing the protection and nurturing of the mindscape, cities can foster a stronger connection between residents and their surroundings, resulting in a more vibrant and resilient urban fabric.

Places and cities hold hidden potential within their context. In urban fabrics, the landscape and its narratives are perceived differently compared to other places. It is a reading derived from a text encompassing values beyond the physical spaces experienced by individuals in other environments. Therefore, the landscape within historical fabrics is a narrative that tells the city's story through its physical manifestation, encompassing collective memories over time, dynamic social activities in different periods, and varied meanings associated with the city's physical structure and essence. The city's story and the human experience of place in different historical periods are narrated through the landscape.

As a new perspective to the revitalization and reconstruction of urban fabrics, the landscape approach and mindscape can contribute to preserving the tangible and intangible values of these places and ensuring the continuity of their identity.

4.2.3 Narrative Landscape

The discussion of narrative clarification initially necessitates defining the concept of narrative. From Leman's perspective, the narrative is a descriptive account of what has happened, understood through a chain of events. Accordingly, a narrative is structured based on the sequence of "this event, then that event" (LeMone et al., 2023). The narrative has been the oldest and most effective means of conveying meaning throughout the history of civilization. The inherent mechanism of the narrative is the representation of natural spaces or the creation of imaginary spaces and situations in which story characters experience life in a way that imparts a lesson to the story's audience.

Constructed space also contains the narration of the ups and downs of individuals' and communities' lives, which they have recounted for future generations and, on a larger scale, transcending the constraints of time and place for all members of society. Therefore, encountering space, particularly a specific place, is a narrative reading in which both the obvious and the hidden teachings of life and its highs and lows are concealed, and understanding it depends on the level of the audience's comprehension and interpretation of the work. The view of any place is a narrative text.

As a narrative text, the landscape is a temporal and contemporary nature and a process resulting from the interpretation and recreation of the text's meaning. Reading the landscape as a narrative text becomes a suitable product of relations interpreted into intertextual relationships. Intertextuality involves connecting layers of one phenomenon with the layers of other phenomena. This leads to developing the landscape's cognitive and social beauty codes as a text in two types of synchronous and diachronic relationships (Raeesi et al., 2023). The reader, with the help of the horizon of semantic connotations, their contemporary knowledge, and intertextual relationships, which means the ties this text has with the reader's previous readings and understanding, engages in the reading of the text. Thus, the text has various subjects that influence its interpretation and open the way for multiple meanings within the text (Rahimi et al., 2021). Therefore, the narrative text of the landscape is influenced by intertextual subjects both in its creation and reading, and the audience engages in the reading through a mental process. Based on this, the landscape results from continually recreating the text in a fresh experience and a different journey constantly renewed among the cognitive horizons of the audience and the text.

Understanding and reading the landscape in historical city fabrics as a form of narrative landscape are also influenced by intertextual subjects. It relies on being present in the city and the mentioned fabrics to identify their various layers. Of course, historical presuppositions, judgments, and personal and collective memories of the audience affect the reading and understanding of the landscape narrative in these places.

Given the provided definition and clarification, the landscape, through its tangible, mental, and metaphorical attributes, narrates the meaning of a place in the form of emotions, activities, and social engagements. As a link between humans and place, the landscape is a narrative that produces, reproduces, and protects another level of human existence, filled with their desires, ideals, beliefs, and myths. Hence, protecting the narrative of the landscape is of particular importance in the process of interventions and occupancies that occur in a place. Any improper substitution or coexistence can influence the narrative of the place and, consequently, affect the connection with the place, the sense of place, and, ultimately, the sense of belonging to the place. This matter is of paramount importance in historical city fabrics, which is the main subject of this paper.

4.3 Case Studies Showcasing Thriving Landscape Approaches in Historical Contexts

4.3.1 Bronzeville: History and Denomination

Bronzeville zone, located on the south side of Chicago, became an established neighborhood around the turn of the twentieth century. Before its more official naming, the media referred to the Bronzeville neighborhood and adjacent areas using derisive names such as the "black belt," "black ghetto," and even more appalling names such as "darkie town." This poor representation on the part of the media coincided with a relatively recent surge in the Black American population in Chicago due to the Great Migration from the southern regions of the United States. The Great Migration began in 1916, and by 1920, the number of Black residents in Chicago had surpassed 100,000. To redirect the community's nominal identity, James J. Gentry, an editor for Anthony Overton, proposed "Bronzeville" as a less derisive name for the small but impactful neighborhood. Thus, the word "Bronzeville" represents not only the community's residents but also resistance and a community's efforts to define itself on its terms (see Fig. 4.4).

Fig. 4.4 Bronzeville past and present. (Source: Graphic courtesy segregation by design team)

Bronzeville can be considered the dream place of African American history. With the beginning of the migration of the first group of Africans living in South America to Chicago in 1779, formerly enslaved people after the American Civil War, in search of a better life, took their way to North America. From the 1940s onward, adventurous Blacks searching for work moved to Chicago and settled in the southern part of the city center. This neighborhood was hit hard by the great economic recession of the United States. However, the worst enemy of the area was the discriminatory local government of Chicago, which unofficially forbade them from renting and buying outside the "black belt" (Hasani Mianroudi et al., 2018).

Bronzeville experienced its cultural renaissance between the 1910s and 1940s. After the 1940s, the neighborhood became increasingly impacted by persistent segregation, with many Black residents unable to find housing anywhere other than the residential neighborhoods that were already primarily Black. In recent years, there have been efforts to revitalize the Bronzeville neighborhood in honor of its previously rich and culturally impactful history. Led by Black entrepreneurs, historical homes are being restored, while new housing also adds to the community. New restaurants and businesses—many of them Black-owned—are beginning to flourish in the community, adding to the sense that the Bronzeville neighborhood is being restored to its former position as a cultural powerhouse (Bones et al., 2013).

Major cities like Chicago or New York in the 2000s were involved with many problems, such as sleeping streets, murder, prostitution, theft, and delinquency. The revitalization of the metropolitan center of Chicago is one of the urban revitalization projects that began in 1994 in the form of the document "Strategic Development Plan for the South of the City Center: Rebuilding Burnsville and Living Around the Boulevards." The main goal of this project has been to revitalize this area based on history, collective memory, and the ethnicities of its inhabitants (Ardakani & Oloonabadi, 2011; Grodach, 2010).

4.3.2 *Projects Implemented in Bronzeville: Objectives and Results*

The process of carrying out the project was based on the participation of the neighborhood's residents in all stages, based on the display and protection of their collective and historical memories. With many successes, this project has improved the quality of life in this neighborhood and attracted many people who come there for recreation and sightseeing. Martin Luther King Boulevard (defender of Black rights, who was killed during the protests) was organized as the spine of Bronzeville. It became the connecting factor of this neighborhood with the neighboring communities. This sculpture is a Black figure with a suitcase and old and torn clothes, which shows his uphill and downhill journey from the south to the north with the hope of a bright future, which he puts his hand on (see Fig. 4.5). Hopes are a sign of success and high hopes. The bronze map of the region is another of the works made in this

Fig. 4.5 Installed bronze statues. (Source: The author's)

Fig. 4.6 Bronze map of the neighborhood. (Source: The authors)

project, which shows the map of Bronzeville along with critical historical points, architecture, parks, and schools (see Fig. 4.6). Also, with the proposal of the "Chicago portraits" project by the cultural association of Chicago neighborhoods based on the recommendation of the municipal cultural affairs department to revitalize the body through history, ethnicity, and culture, portraits of people who were the source of services in the neighborhood and until now have not been appreciated, together with their life stories, were presented in a permanent exhibition. The history and ethnic groups of Chicago were also represented with the help of "mosaic art" in urban spaces (Grodach, 2010).

The Bronzeville neighborhood revitalization project has been carried out to preserve and strengthen the neighborhood's collective memories and attract citizens' participation. This project is defined through the definition of various thematic projects, such as the projects "eyes on the walls of the city of Chicago," "mosaic art," "bronze sculpture," "bronze map of the neighborhood," and "23 benches collections" and the program "Who are you?" People have been called to the field of activity in revitalizing the neighborhood, and exciting results have been achieved. It is worth mentioning that the connection of these projects with the collective memories and the spending of money, time, and energy by the citizens in carrying out these projects, on the other hand, increased the "sense of belonging" and strengthened the "total social cohesion" and also justified it (Schlabach, 2012). Common ground has been established among the residents. As a result, it has led to the "enhancement of the cooperative spirit" of the residents of this neighborhood. Another exciting development of this project is using this statue's image as the city council's logo in Bronzeville, which has become the official symbol of the Bronzeville neighborhood. "Bronze statue" received much attention and became a distinguishing feature and character of the Bronzeville neighborhood among the citizens. Therefore, the revitalization project of downtown Chicago (Bronzeville) with the help of projects such as "bronze sculpture," "bronze map," "set of 23 benches," "eyes on the walls of the city of Chicago," and "mosaic art" is successful in redefining the sense of performance (see Fig. 4.7).

By effectively tapping into the collective memories of the residents, the overseer has skillfully crafted a strong sense of place, fostering a sense of community, belonging, and social cohesion. As a result, it has gained widespread acceptance (Schlabach, 2012). It has attracted private sector investment and environmental protection, urban furniture has become a part of people's minds, and the place's attractiveness has become in people's minds. It is safe for tourists. Now, as the weather gets dark, the Bronzeville is full of people, and the presence of local drama, dance, and music brings the citizens to spend their free time. As time changes, Bronzeville is full of people, but the presence of local drama, dance, and music invites citizens

Fig. 4.7 An example of mosaic work discovered in the excavations around the Crown Hall building. (Source: Abel Uribe/Chicago Tribune)

to spend their free time in this neighborhood. The projects carried out by the city administration of Chicago (Martin et al., the construction and installation of a bronze statue, a bronze map, portraits of Chicago, eyes on the walls of the city, and the collection of 23 benches) are consistent with the upgrade, the sense of belonging. So, it has been revitalized by improving the quality in physical-spatial, social, and cultural dimensions. The collection of 23 benches with different names taken from the collective memories of people with different artistic views were designed and built by neighborhood artists. Three bars are untitled. However, the rest of each has a title that talks about the culture and history of the neighborhood. Body, soul, youth, happy couple, Queen Emmys or guitar, a walk around the neighborhood, and Louis and Lil are among the names of the benches (Grodach, 2010).

The program "Who are you?" was designed by the Grand Boulevard neighborhood arts pioneers and is another similar program. Several organizations managed this program. The slogans of this program are as follows: you are colorful (using rainbows to paint the walls), you are dramatic (using young people and teenagers in designing street and popular shows), you have words (teaching various professions to young people and teenagers), means of urban enterprises, you are green (participating young people and teenagers in the preservation and development of green spaces), and you are great (participating young people and teenagers in preserving the city's cultural heritage), which improve the sense of belonging and collective memory (Grodach, 2010). In alignment with the project's objectives and based on the theoretical studies conducted in preceding sections, questionnaires were formulated in two parts for the residents of the target area and experts in the field. These questionnaires aimed to validate and assess the indicators identified in the theoretical studies section and refine and augment them with the research objectives (see Appendix A and Appendix B).

4.4 Conclusion: Theoretical Framework of Landscape Approach in Historic Cities

As mentioned earlier, the landscape is a visual amalgamation of the values of a place, narrating the experiences and collective memories associated with the tangible elements and spaces that have endured from the past (Fig. 4.6). It directly relates the sense of place to a feeling of belonging, evoking nostalgic memories and more. Therefore, within courtyards and historical urban fabrics, the landscape can play a crucial role in the sustainability and revitalization of these areas—landscape—Fig. 4.8.

Architects and their management provide a simultaneous foundation for creating and sustaining meaning within historical contexts. This foundation offers a holistic perspective on historic fabrics, which ultimately leads to their revitalization and rejuvenation.

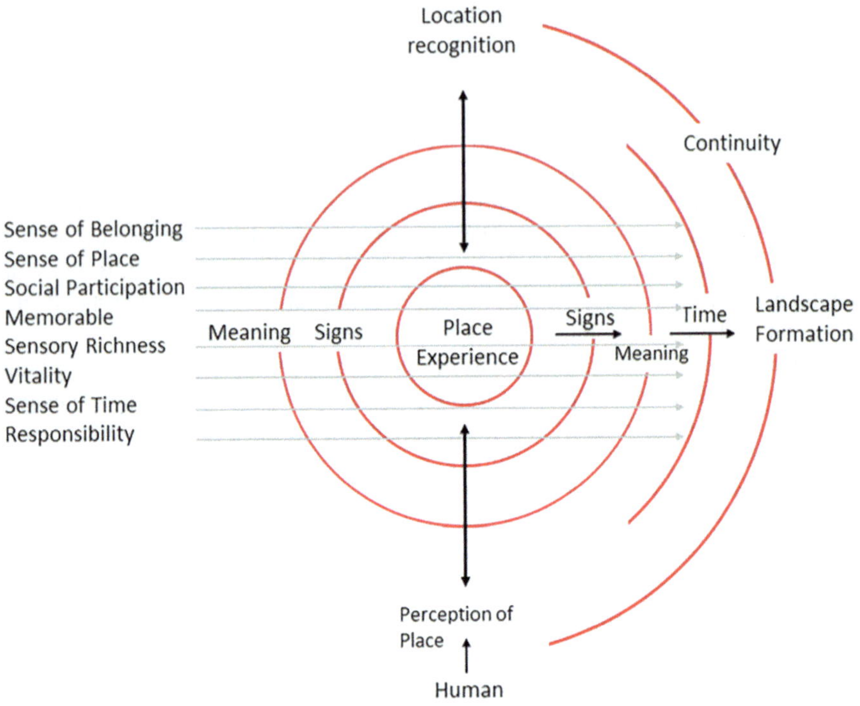

Fig. 4.8 The landscape and the human and place components that are effective in its formation. (Source: Authors)

Historical places and fabrics have hidden potential within their context. The landscape and its narrative are interpreted differently in historical fabrics than elsewhere. This interpretation is derived from a text with transcendent values beyond humans' spatial experiences in other areas. Thus, the landscape in historical fabrics is a text that narrates the collective memories over time, dynamic social activities in different periods, and varying meanings at each moment, along with the physical and tangible story of the city. The city's story and the human experience of place in different historical periods are narrated through the landscape. In continuation, considering the conceptual model of the research, the neighborhood behind the dome of Isfahan is examined as a case study in terms of landscape management, collective memories, placemaking, and intervention methods. This neighborhood has changed its physical structure and social fabric from the Safavid period until now, and it possesses potential relevant to the present research background. The presence of historical elements and characteristics in the neighborhood fabric and its vicinity amplifies the importance of this neighborhood. However, interventions and encroachments on the fabric have taken place regardless of landscape values, which will be addressed further in Fig. 4.9.

As a new perspective in the revitalization and regeneration of historical fabrics, the landscape approach can contribute to preserving the tangible and intangible

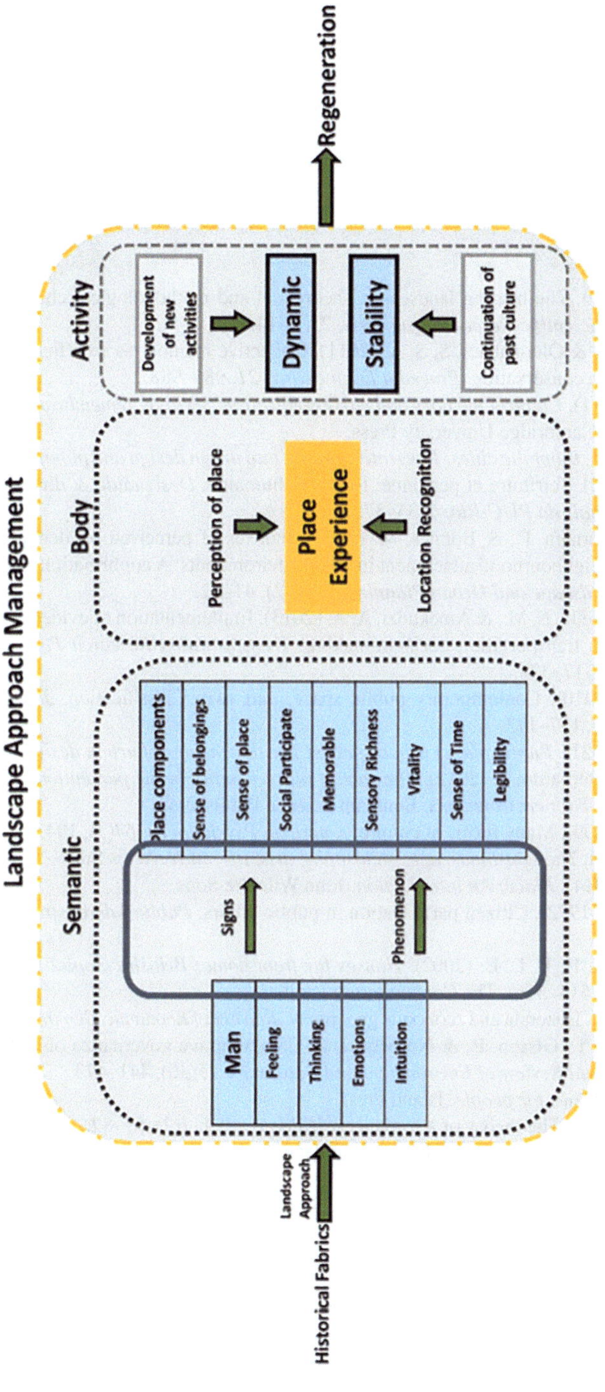

Fig. 4.9 Landscape management model in landscape recreation in historical contexts. (Source: Authors)

values of these places and the continuity of their identity. The landscape can maintain and express the city's historical narrative on a larger scale. The landscape of historical sites is a rich textual environment, and any intervention, alteration, or interaction with it is only possible by reading and understanding the relevant narrative (Khorasgani et al., 2023).

References

Ansari, M. (2017). The healing landscape: Theoretical and methodological challenges. *Art and Architecture Scientific-Research Quarterly, 25*, 5–14.

Ardakani, M. K., & Oloonabadi, S. S. A. (2011). Collective memory as an efficient agent in sustainable urban conservation. *Procedia Engineering, 21*, 985–988.

Assmann, J. (2011). *Cultural memory and early civilization: Writing, remembrance, and political imagination*. Cambridge University Press.

Beatley, T. (2011). *Biophilic cities: Integrating nature into urban design and planning*. Island Press.

Berque, A. (2010). Territoire et personne: l'identité humaine. *Desigualde & diversidade–Rivista de viesna socials da PUC-Rio, 6*, 35–37.

Bonaiuto, M., Fornara, F., & Bonnes, M. (2003). Indexes of perceived residential environment quality and neighbourhood attachment in urban environments: A confirmation study on the city of Rome. *Landscape and Urban Planning, 65*(1–2), 41–52.

Bones, E. J., Barrella, E. M., & Amekudzi, A. A. (2013). Implementation of evidence-based design approaches in transportation decision-making. *Transportation Research Part A: Policy and Practice, 49*, 317–328.

Carmona, M. (2010). Contemporary public space, part two: Classification. *Journal of Urban Design, 15*(2), 157–173.

Carmona, M. (2021). *Public places urban spaces: The dimensions of urban design*. Routledge.

Casakin, H., & Bernardo, F. (2012). *The role of place identity in the perception, understanding, and design of built environments*. Bentham Science Publishers.

Cohen, A. B. (2009). Many forms of culture. *American Psychologist, 64*(3), 194.

Collot, M. (2011). *Thought-landscape, philosophy, arts, literature*. Actes Sud.

Cresswell, T. (2014). *Place: An introduction*. John Wiley & Sons.

Cunningham, J. (1972). Citizen participation in public affairs. *Public Administration Review, 32*, 589–602.

Dominy, M. I. C. H. È. L. E. (2002). *Houses far from home: British colonial space in the new hebrides* (pp. 514–516). The Contemporary Pacific.

Florida, R. (2002). Bohemia and economic geography. *Journal of Economic Geography, 2*(1), 55–71.

Folke, C., Hahn, T., Olsson, P., & Norberg, J. (2005). Adaptive governance of social-ecological systems. *Annual Review of Environment and Resources, 15*(30), 441–473.

Gehl, J. (2013). *Cities for people*. Island Press.

Gibson, J. J. (1977). The theory of affordances. *Hilldale, USA, 1*(2), 67–82.

Grodach, C. (2010). Art spaces, public space, and the link to community development. *Community Development Journal, 45*(4), 474–493.

Halbwachs, M. (1992). *On collective memory*. University of Chicago Press.

Hasani Mianroudi, N., Majedi, H., Saeide Zarabadi, Z. S., & Ziari, Y. (2018). Exploring concept of collective memory and its retrieval in urban areas with semiotic approach (Case study: Hasan-Abad square). *The Monthly Scientific Journal of Bagh-e Nazar, 14*(56), 17–32.

Hesami, S., Sedaghati, A., & Oskouei, B. (2022). Analysis of human interaction with the environment in terms of the effects of perception and behavior on the place attain cement (A case study of Mehr Mokrian Housing Complex in Mahabad). *Urban Design Discourse-a Review of Conte Literatures and Theories, 3*(1), 39–57.

Heydari, M. (2020). *Reading the archaeological landscape (reconstructing perceptual and ecological layers)* (Unpublished Master thesis in Landscape Architecture). Faculty of Art, Tarbiat Modares University, Tehran, Iran.

Hou, J. (2010). *Insurgent public space: Guerrilla urbanism and the remaking of contemporary cities*. Routledge.

Huyssen, A. (2003). *Present pasts: Urban palimpsests and the politics of memory*. Stanford University Press.

Innes, J. E., & Booher, D. E. (2003). *Planning with complexity | An Introduction to collaborative rationality*(taylorfrancis.com).

Kaltenborn, B. P. (1998). Effects of the sense of place on responses to environmental impacts: A study among residents in Svalbard in the Norwegian high Arctic. *Applied Geography, 18*(2), 169–189.

Kashi, H., & Bonyadi, N. (2013). Stating the model of identity of place-sense of place and surveying its constituents* case study: Pedestrian passage of Shahre Rey. *Journal of Fine Arts: Architecture & Urban Planning, 18*(3), 43–52. https://doi.org/10.22059/jfaup.2013.51317

Khorasgani, A. M., & Haghighatbin, M. (2023). Regeneration of historic cities: Reflections of its evolution towards a landscape approach. *ISVS e-Journal, 10*(5), 60–78.

Khorasgani, A. M., Villalobos, M. H., & Eskandar, G. A. (2023). Sustaining historic cities: An approach using the ideas of landscape and place. *ISVS e-Journal, 10*(1), 320–332.

LeMone, M. A., Ikeda, K., & Angevine, W. M. (2023). Daytime convective boundary-layer evolution on three fairweather days in CASES-97. *Boundary-Layer Meteorology, 187*(3), 527–565.

Lewicka, M. (2011). Place attachment: How far have we come in the last 40 years?. *Journal of Environmental Psychology, 31*(3), 207–230.

Lynch, K. (1972). *The openness of open space, arts of environment*. Aidan Ellis.

Lynch, K. (2013). "The Image of the Environment" and "The City Image and Its Elements": From The Image of the City (1960). In *The urban design reader* (pp. 125–138). Routledge.

Mansouri, S. (2005). An introduction to Landscape architecture identification. *The Monthly Scientific Journal of Bagh-e Nazar, 1*(2), 69–78.

Mansouri, M. (2008). Landscape, place, history. *The Monthly Scientific Journal of Bagh-e Nazar, 5*(9). 81–90.

Mansouri, S. A., & Farzin, A. (2016). *A landscaped approach in urban design, a methodology for urban space designs*. Tehran: Nazar Research Center.

Manzo, L., & Devine-Wright, P. (2013). *Place attachment: Advances in theory, methods, and applications*. Routledge.

McAdams, D. P. (1989). The development of a narrative identity. In *Personality psychology: Recent trends and emerging directions* (pp. 160–174). New York, NY: Springer US.

Mirmoqtadaei, M. (2004). Standards for discovering and evaluating urban physical identity. *Journal of Final Arts Faculty, 11*(1), 17–26.

Mohammadi, H. (2010). *Citizen participation in urban planning and management: The case of Iran, Shiraz City, Saadi community*. Kassel University Press GmbH.

Norberg-Schulz, C. (1981). Chicago: Vision and image. In M. Casari & V. Pavan (Eds.), *New Chicago architecture* (pp. 57–70).

Norberg-Schulz, C. (2019). Genius loci: Towards a phenomenology of architecture (1979). *Historic Cities: Issues in Urban Conservation, 8*, 31.

Parkins, W., & Craig, G. (2006). Slow living. .

Proshansky, H. M., Fabian, A. K., & Kaminoff, R. (1983). Place-identity: Physical world socialization of the self. *Journal of Environmental Psychology, 3*(1), 57–83.

Raeesi, R., Sahebjamnia, N., & Mansouri, S. A. (2023). The synergistic effect of operational research and big data analytics in greening container terminal operations: A review and future directions. European *Journal of Operational Research, 310*(3), 943–973.

Rahimi, E., Barghjelveh, S., & Dong, P. (2021). Quantifying how urban landscape heterogeneity affects land surface temperature at multiple scales. *Journal of Ecology and Environment, 45*, 1–13.

Rapoport, A. (1982). Urban design and human systems: On relating buildings to the urban fabric. In *Human and energy factors in urban planning: A systems approach*. Springer.
Rashid Kolvir, H., Abolghasemi, A., Akbari, H., & Shoaib, K. M. (2019). The role of physical characteristics in predicting Rasht's residential complex residents' psychological and social indicators. *Motaleate Shahri, 8*(32), 65–76.
Relph, E. (1976). *Place and placelessness* (Vol. 67). Pion.
Ryan, R. L. (2006). The role of place attachment in sustaining urban parks. *The human metropolis: People and nature in the 21st-century city* [complete book] (p. 61). University of Massachusetts Press
Schlabach, E. S. (2012). *Along the streets of Bronzeville: Black Chicago's literary landscape*. University of Illinois Press.
Shamai, S. (1991). Sense of place: An empirical measurement. *Geoforum, 22*(3), 347–358.
Smith, M. D. (2011). The ecological role of climate extremes: current understanding and future prospects. *Journal of Ecology, 99*(3), 651–655.
Stedman, R. C. (2002). Toward a social psychology of place: Predicting behavior from place-based cognitions, attitude, and identity. *Environment and Behavior, 34*(5), 561–581.
Tuan, Y. F. (1993). *It is passing strange and wonderful: Aesthetics, nature, and culture*. Island Press.
Ward, K., Lauf, S., Kleinschmit, B., & Endlicher, W. (2016). Heat waves and urban heat islands in Europe: A review of relevant drivers. *Science of the Total Environment, 569*, 527–539.
Yorkey, R. (1979). Longman Dictionary of Current English.

Chapter 5
Data Science for Historic City Analysis

5.1 Introduction to Data Science

As a multidisciplinary field, data science integrates scientific methods, processes, algorithms, and systems to extract insights from diverse data sources. In the context of historic city analysis, data science becomes a powerful tool for unraveling the intricate layers of a city's past. It comprehensively explains urban development, societal changes, and cultural evolution. The interdisciplinary nature of data science allows for integrating techniques from statistics, computer science, and domain-specific expertise in history and urban studies (Provost & Fawcett, 2013). Through data science, researchers and historians gain the tools to uncover hidden patterns, validate hypotheses, and derive meaningful insights from the wealth of historical records (Cao, 2017).

By integrating data science into urban planning, we are not just adding a new tool but transforming the field. This transformation offers valuable insights into analyzing historic cities, opening new avenues for understanding and preserving our urban heritage. This essay delves into the significance and potential applications of data science methodologies in urban planning, particularly in the delicate balance between historical preservation and contemporary development.

Data science applied to historic city analysis involves leveraging machine learning, statistical analysis, and data visualization techniques to uncover hidden patterns in historical datasets. For example, using natural language processing, historians can analyze vast amounts of textual data from historical documents, providing insights into societal trends, cultural shifts, and economic changes.

A landscape approach involves understanding the interconnectedness of various elements in a city, considering both natural and built environments. Data science, when applied with a landscape approach, enables the comprehensive analysis of historic landscapes, including changes in land use, ecological systems, and the built environment (Turner, 2015) utilizing remote sensing data to analyze changes in

A. Moazzeni Khorasgani, *Using Data Science and Landscape Approach to Sustain Historic Cities*, https://doi.org/10.1007/978-3-031-68161-5_5

urban green spaces over time, helping historians understand the evolution of parks, gardens, and recreational areas within the city.

5.2 Data Collection Methods

Efficient data collection is paramount in leveraging data science for historic city analysis. Traditional methods, including surveys and archival research, are now complemented by modern technologies such as Geographic Information System (GIS), remote sensing, and crowdsourced data. These approaches provide a comprehensive understanding of a city's historical evolution and empower planners with real-time data for decision-making.

5.2.1 Historical Records

Delving into archives, documents, and records provides a treasure trove of historical data. This method involves meticulously examining primary sources, such as governmental documents, newspapers, and manuscripts (Longley et al., 2015). Analyzing census records, historical newspapers, and governmental documents can provide details about population growth, economic activities, and political events.

Analyzing digitized historical newspapers using text mining techniques allows for extracting information on events, sentiments, and public opinions during specific periods. The holistic view of the landscape to the historical tissues and cities and the continuation of this attitude to the historical sites can create a comprehensive approach to data collection. The comprehensive approach of collecting objective and subjective data records the tangible and intangible heritage.

5.2.2 GIS (Geographic Information System)

Utilizing GIS technology helps collect and analyze spatial data, enabling maps that showcase a city's geographical evolution over time (Anderson et al., 1976). Mapping changes in land use over time using GIS helps visualize how the urban landscape has evolved, aiding in identifying trends and development patterns. Integrating historical maps into GIS systems enables researchers to visualize changes in urban infrastructure, land use, and population density over time (Talbot et al., 2018). They overlay historical maps with ecological data to analyze how urbanization has affected biodiversity and ecosystems within the city.

5.2.3 Sensor Data

Deploying sensors for environmental monitoring or infrastructure usage yields real-time or historical data, contributing valuable insights into the city's development. Sensors deployed to monitor air quality or traffic flow can offer insights into environmental conditions and infrastructure usage during different historical periods. Historical weather data collected through sensors can provide insights into climate changes, influencing the city's growth and development (Hannak & Beigl, 2017). This system can be integrated with IoT—the Internet of Things—and developed to be more efficient.

5.2.4 Crowdsourcing

Engaging the community in data collection fosters a collaborative approach, tapping into the collective knowledge of residents, historians, and enthusiasts (Longley et al., 2015). Engaging the public in digitizing and transcribing historical documents through online platforms like crowdsourced transcription projects facilitates the collective effort in preserving and digitizing valuable historical information. Platforms like Zooniverse enable crowdsourced transcription of historical documents, engaging volunteers in preserving and digitizing valuable data (Simpson et al., 2014) and involving the community in identifying and documenting historical landscape features through crowd-contributed photographs and oral histories.

Another feature of the landscape approach is the creation of public and social participation and the narration of collective memories. In this direction, the synergy between landscape and data science can lead to continuing sustainability goals in historical cities.

5.3 Data Analysis Techniques and Models

5.3.1 Temporal Analysis

Examining temporal patterns in historical data reveals trends and cyclical changes, offering insights into the city's evolution over different periods (James et al., 2013). Studying changes in economic indicators over decades allows for identifying economic cycles and the impact of historical events on the city's economic landscape. Time series analysis of historical economic data can reveal long-term economic trends and the impact of historical events on the city's financial landscape (Hyndman & Athanasopoulos, 2018). We are studying changes in land cover over time to analyze the impact of urbanization on natural landscapes, such as wetlands or forests.

5.3.2 Spatial Analysis

Analyzing the spatial distribution of historical data through techniques such as spatial autocorrelation helps uncover patterns and relationships between different locations. Spatial autocorrelation can reveal clusters of historical events or changes in population density, indicating areas of significant historical importance or growth. Using hotspot analysis in GIS helps identify clusters of historical events, guiding historians to focus on specific regions with unique historical significance (Anselin et al., 2006). Applying spatial analysis to identify historical green space distribution patterns and access reveals socio-environmental disparities.

5.3.3 Machine Learning Models

Applying machine learning models enables predictive analysis, allowing forecasting of future trends based on historical patterns (Barabási, 2016). Implementing regression models on historical datasets can help predict future trends based on patterns identified in past data, aiding in informed decision-making. Applying predictive models to historical data can help forecast urban growth patterns, aiding city planners in making informed decisions for future development (Khandelwal & Fan, 2020).

Deploying environmental sensors to monitor changes in air quality and green coverage contributes to a holistic understanding of the city's historical landscape health. Machine learning algorithms are utilized to predict the historical expansion of urban areas and assess their impact on the natural landscape.

5.3.4 Network Analysis

Investigating the interconnectedness of historical events or societal elements through network analysis provides a holistic view of the city's development. Analyzing historical social networks, such as connections between influential figures or organizations, provides insights into the city's development dynamics. Analyzing historical social networks through network analysis techniques reveals influential figures and organizations, providing insights into power structures and societal dynamics (Scott, 2017). They analyze the interconnectedness of historic landscapes, including transportation routes and green corridors, to understand their role in shaping the city's development.

5.4 Make Decisions Based on Results and Monitoring

5.4.1 Policy Recommendations

Informed by data-driven insights, policymakers can formulate effective strategies for preserving or developing historical areas, ensuring a balance between conservation and progress (Bertaud, 2018). Using data insights, policymakers can prioritize historic preservation efforts for areas with cultural significance, while balancing the need for urban development. Historical crime data analysis can inform public safety policies, allowing policymakers to allocate resources effectively to areas with a history of criminal activity (Mohler et al., 2011). They are implementing policies for green infrastructure development based on historical landscape analysis to enhance ecological resilience and biodiversity.

5.4.2 Urban Planning

Data science contributes to urban planning by providing insights that influence the design and development of urban spaces, considering historical significance and future needs.

Integrating historical data into urban planning ensures that new developments consider the preservation of historical landmarks and maintain the city's unique identity (Glaeser, 2011). Incorporating historical preservation zones into urban planning regulations helps maintain the city's historical identity, while accommodating modern development needs (Tyler & Moench, 2012). Integrating historical landscape features into urban planning guidelines would incorporate green spaces in new developments.

5.4.3 Monitoring Changes

Continuous monitoring and data analysis allow for adapting strategies in response to ongoing developments, ensuring that decisions align with the dynamic nature of the city. Continuous monitoring of demographic shifts and economic indicators enables quick adaptation of policies to address emerging challenges and opportunities in the city. Real-time monitoring of traffic patterns and infrastructure usage can guide immediate adjustments in urban planning strategies to address emerging challenges (Zhong et al., 2016). Real-time monitoring of changes in urban vegetation through satellite imagery would adopt conservation strategies and address emerging environmental challenges.

5.5 Conclusion

Extracting meaningful insights from collected data is critical to utilizing data science to understand urban dynamics. Machine learning algorithms, statistical analyses, and predictive modeling are among the techniques employed by urban planners to discern patterns, forecast trends, and optimize resource allocation for historic city preservation and development.

Case studies best illustrate real-world applications of data science in historic city analysis. Cities like Barcelona, Rome, and Kyoto have successfully implemented data-driven approaches to revitalize historic areas while maintaining cultural integrity. For instance, Barcelona utilized data analytics to enhance tourism experiences. Rome optimized restoration efforts through data analysis, and Kyoto managed the delicate balance between cultural heritage and modern development using data science.

Despite the advantages, integrating data science in historic city analysis poses challenges that need addressing. Data privacy, ethical considerations, and algorithmic biases require careful attention. Interdisciplinary collaboration among data scientists, urban planners, historians, and local communities is essential for a holistic and culturally sensitive approach to historic city analysis.

Looking ahead, the future of data science in historic city analysis appears. Advances in technology, coupled with a growing awareness of the importance of preserving cultural heritage, will likely lead to more innovative and effective solutions. Ongoing collaboration between urban planning and data science will continue to shape sustainable development and conservation strategies for historic cities.

In conclusion, integrating data science into urban planning, particularly in analyzing historic cities, is a transformative endeavor. Through advancements in data collection, analysis techniques, and insightful case studies, data science is a guiding force for making informed decisions that balance preserving and developing historic urban spaces. As we navigate the complexities of urban development, data science stands as a beacon, guiding us toward a harmonious coexistence of the past and the future.

References

Anderson, J. R., Hardy, E. E., Roach, J. T., & Witmer, R. E. (1976). *A land use and land cover classification system with remote sensor data* (p. 964). U.S. Geological Survey Professional Paper.

Anselin, L., Syabri, I., & Kho, Y. (2006). GeoDa: An introduction to spatial data analysis. *Geographical Analysis, 38*(1), 5–22.

Barabási, A. L. (2016). *Network Science*. Cambridge University Press.

Bertaud, A. (2018). *Order without design: How markets shape cities*. MIT Press.

Cao, L. (2017). *Data science: An introduction*. CRC Press.

Glaeser, E. L. (2011). *Triumph of the city: How our greatest invention makes us richer, smarter, greener, healthier, and happier*. Penguin.

Hannak, A., & Beigl, M. (2017). Sensing the past: Enabling large-scale historical data research through smartphones. *Journal of Ambient Intelligence and Humanized Computing, 8*(6), 799–809.

Hyndman, R. J., & Athanasopoulos, G. (2018). Forecasting: principles and practice. *OTexts.*

James, G., Witten, D., Hastie, T., & Tibshirani, R. (2013). *An introduction to statistical learning*. Springer.

Khandelwal, S., & Fan, W. (2020). Urban growth prediction using machine learning: A case study of Bangalore, India. *ISPRS International Journal of Geo-Information, 9*(9), 534.

Longley, P., Goodchild, M. F., Maguire, D. J., & Rhind, D. W. (2015). *Geographic information science and systems*. Wiley.

Mohler, G., Short, M. B., Malinowski, S., Johnson, M., Tita, G., Bertozzi, A. L., & Brantingham, P. J. (2011). Randomized controlled field trials of predictive policing. *Journal of the American Statistical Association, 106*(493), 735–745.

Provost, F., & Fawcett, T. (2013). Data science for business. O'Reilly Media.

Scott, J. C. (2017). *Against the grain: A deep history of the earliest states*. Yale University Press.

Simpson, R., Page, R., De Roure, D., & Crawford, T. (2014). Zooniverse: Observing the world's largest citizen science platform. In *Proceedings of the 23rd international conference on the World Wide Web* (pp. 1049–1054).

Talbot, R., Fairbairn, D., & Martin, G. (2018). *Geographical information systems: Principles and applications*. Wiley.

Turner, W., Spector, S., Gardiner, N., Fladeland, M., Sterling, E., & Steininger, M. (2015). Free and open-access satellite data are essential to biodiversity conservation. *Biological Conservation, 182*, 173–176.

Tyler, S., & Moench, M. (2012). A framework for urban climate resilience. *Climate and Development, 4*(4), 311–326.

Zhong, C., An, L., Gao, J., & Li, X. (2016). Real-time urban monitoring in the era of big data. *GIScience & Remote Sensing, 53*(4), 465–484.

Chapter 6
Sustainable Development Strategies for Historic Cities

6.1 Introduction: Sustainable Regeneration and Design Principles

Numerous studies, designs, and research have been conducted to pursue sustainable revitalization of historical fabrics. The proposed solutions can be broadly categorized into two groups. The first group focuses on the physical aspects of historical fabrics, addressing issues related to buildings and their physical interactions. In contrast, the second group offers behavioral, sociocultural, and semantic studies regarding historical fabrics. Both solution categories could have been more effective, providing a somewhat discrete and mental perspective on these valuable places.

Hence, the primary concern of this research is how to employ a holistic approach in the preservation and revitalization of the values of historical fabrics and strengthen the collective memory's role as an influential factor in the continuity of these places in redevelopment and revitalization programs.

In this research, after defining the landscape and its relationship with the place and examining the influential indicators in enhancing this connection, an analysis of the landscape characteristics and the places of historical fabrics, along with a review of various theories by experts in this field, was carried out. This is done to identify landscape values and their impact on the identity of historical fabric, aiming to create a general road map for the sustainable regeneration of these places. Today, various approaches and solutions have been proposed regarding the conservation, revitalization, and redevelopment of historical fabrics in cities.

In contemporary architectural research, exploring historical urban fabrics goes beyond cultural, artistic, and historical perspectives. It has transformed into a dynamic, enduring force (Smith & Jones, 2021). Preserving authenticity in historical heritage is now recognized as a critical driver of global cultural diversity (Brown & White, 2020). Historical urban fabrics encompass tangible and intellectual elements, with memories that transcend their temporal origins, forming an integral part

A. Moazzeni Khorasgani, *Using Data Science and Landscape Approach to Sustain Historic Cities*, https://doi.org/10.1007/978-3-031-68161-5_6

of society's identity and social fabric (Anderson & Wilson, 2019). These values manifest diversely within urban environments, further enhancing their significance and prominence (Thomas et al., 2022).

Neglecting the semifictional identity and the inhabitants' cultural and everyday life requirements in restoring historical urban fabrics can result in their transformation into lifeless spaces, merely retaining their historic facades. Therefore, it is imperative to emphasize the continuum of the past and present, treating the city as an interconnected historical-urban fabric rather than a collection of isolated, disjointed historical elements. This holistic approach views the city as a living, dynamic organism (Jackson, 2018).

In this context, the central question arises: how can a landscape-oriented approach effectively contribute to the sustainability of historical urban fabrics? What specific tangible and intellectual indicators within the urban landscape can play a pivotal role in successfully revitalizing these areas? It becomes apparent that a focus on landscape and its inherent values within historical urban fabrics holds immense potential, and this matter will be explored in depth in the subsequent sections of this research.

The findings of this research can be categorized into two main groups. The first category comprises insights from analyzing relevant articles and research studies. The second category encompasses the recommendations and guidelines provided by organizations associated with regenerating historic cities, such as UNESCO.

6.2 Landscape Approach Criteria in Research and Development

The perspective of landscape professionals encompasses a holistic view of all the influential factors and elements in the historic urban fabric. This perspective is aimed at achieving comprehensive continuity and regeneration in historical contexts. Landscape professionals play a crucial role in identifying and defining the signs and phenomena in the natural environment, enabling a profound understanding of human sensory and perceptual aspects. This understanding, in turn, contributes to the effective management and regeneration of historical urban fabrics. In essence, landscape plays an active role as a mediator between humans and their surroundings, creating meaningful elements in historical contexts. These elements encompass emotions, perceptions, thoughts, and insights, which, through the lens of landscape professionals, give rise to factors such as a sense of belonging, personal and collective nostalgia, a feeling of participation, a connection to time and place, and an enhancement of sensory richness, thereby enriching social life and ensuring the dynamism and stability of historical urban fabrics. Each of these elements represents a dynamic and stabilizing force within historical contexts, and their management guarantees the sustained existence of these living ecosystems.

The notion of "landscape" in contemporary architecture is a modern and connective system that profoundly influences one's relationship with a place, eliciting qualities such as a sense of place and a feeling of belonging. The sense of belonging plays a pivotal role in the continuity of a place. Historic places inherently possess a rich narrative conveyed and interpreted through the landscape. Reading the narrative of the landscape facilitates human connection with the place, contributing significantly to the cultivation of a sense of belonging, collective memory, and social engagement factors in the process of urban regeneration. A landscape perspective offers a holistic view of all factors and elements affecting historical contexts to achieve the continuity of identity and foster the comprehensive engagement of citizens with these spaces during the regeneration process. With its narrative capacity, landscape delves deeper into the introduction of a place and actively strengthens the sense of place and belonging (Moazzeni Khorasgani & Villalobos, 2023).

Historic sites are fundamentally rich texts, and any form of intervention, alteration, or interaction with them without proper interpretation and comprehension of their narratives can lead to misreading and a failure to grasp the historical account. Therefore, incorporating a landscape approach in revitalizing and preserving historical contexts can facilitate the correct interpretation of the landscape narrative and the perception of the historical account within these historical settings. In this regard, embracing landscape methodologies is imperative for the sustainable conservation and enhancement of historical cities. To present a conceptual model for research into the sustainable regeneration of historical contexts, it is essential to conduct a comprehensive analysis and understanding of the issues and opportunities related to these historical contexts. Gathering data and completing information, including historical sources and residents' collective memories, can be paramount in this endeavor. A comprehensive understanding of the current issues (physical, functional, and semantic) related to these contexts and examining all tangible and intangible aspects pave the way for subsequent steps in the process (Khakzand & Jafari Khaledi, 2023).

The perspective of landscape architects entails a holistic examination of all influential factors in the historic urban fabric to achieve comprehensive continuity and regeneration of historical areas. Landscape architects play a pivotal role in defining and elucidating the signs and phenomena within ecological settings to ultimately contribute to managing sensory and perceptual aspects of human experiences for the sustained preservation and revitalization of the historical fabric (Eskandar et al., 2022) (Fig. 6.1).

In essence, landscape architecture serves as the intermediary between human beings and their environment, catalyzing the creation of meaningful components within historical urban settings. These components encompass emotions, perceptions, thoughts, and insights that, through the lens of landscape architecture, give rise to factors such as a sense of belonging, collective and personal nostalgia, a feeling of participation, awareness of time and place, and heightened sensory richness, all enriching social life. These components contribute to the dynamism and stability of the historic urban fabric, ensuring the continuous vitality of these living ecosystems through their effective management.

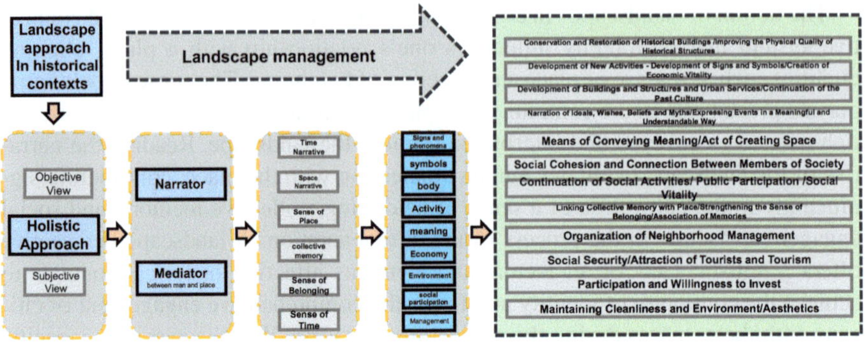

Fig. 6.1 Landscape indicators in historical contexts based on library sources. (Source: Author)

In general, the theoretical findings of library resources can be divided into the following general principles.

Preserving the historical elements of a city is crucial. This includes maintaining physical structures, symbols, and objects representing the city's history. Historical preservation ensures the connection of future generations with their roots and helps understand the city's evolution over time. It also contributes to preserving the city's unique identity. Beyond physical structures, heritage includes nonmaterial aspects such as traditions, customs, language, and culture. Preserving heritage and culture is vital to maintaining the authenticity of a city. This includes preserving traditions, rituals, and practices passed down from generation to generation, ensuring that the city's cultural identity remains vibrant and continually influences it. Indigenous knowledge and traditional methods are valuable resources for sustainable urban development. Local wisdom and traditional methods provide sustainable solutions to contemporary challenges. Promoting and incorporating these methods into modern urban planning can enhance the resilience and sustainability of the city (Tyler et al., 2018).

Creating and disseminating historical stories is essential. This includes documenting and sharing stories, events, and individuals who have shaped the city's history. Compelling storytelling can deepen residents' and visitors' understanding of the city's historical heritage. Heritage tourism encourages visitors to explore a city's historical and cultural elements. Promoting heritage tourism generates economic benefits and increases awareness of the importance of preserving heritage sites. It can also create job opportunities for residents. Collaborating with the local community is crucial for heritage preservation (Ashworth & Tunbridge, 2000). Involving residents in decision-making and seeking their input on heritage-related projects instills a sense of ownership and pride. Community participation can lead to innovative approaches to heritage preservation. Education and increasing awareness about a city's history and heritage are fundamental. Schools, museums, and cultural institutions significantly educate residents and visitors about the city's past. Public campaigns and cultural events can enhance the appreciation and valuation of heritage (Lowenthal, 1998).

Achieving social justice in historical cities is crucial. This involves ensuring all residents have equal access to resources and opportunities regardless of their economic and social backgrounds. Programs and policies aimed at social justice can reduce income and wealth disparities and create a more equitable urban environment. Respecting and preserving human rights are essential for restoring historical cities. Human rights encompass various aspects, such as adequate housing, education, and cultural participation. Upholding these rights ensures that residents can live with dignity and be free to engage in cultural and social activities (Duignan, 2023). Promoting social welfare programs in historical cities can improve residents' overall quality of life. Access to healthcare, education, and social services is essential to ensure the community's well-being. Investing in social welfare contributes to the overall improvement of the city. Promoting economic growth and prosperity in historical cities is essential. This can be achieved by encouraging entrepreneurship, supporting local businesses, and attracting investments. A thriving economy can create job opportunities and generate income for further development. Implementing strategies to reduce poverty is vital. This includes targeted programs to support marginalized communities, provide vocational training, and ensure access to necessities. Reducing the poverty rate strengthens the social framework of the city. Promoting ethical business practices is necessary for sustainable development. Historical cities can enforce regulations and incentives to encourage responsible business behavior. Ethical business practices contribute to the preservation of cultural heritage and environmental sustainability. Promoting social innovation means finding creative solutions to urban challenges. This can include programs that strengthen community participation, promote cultural preservation, and develop communal spaces. Social innovation contributes to a more resilient and inclusive city (Henning-Smith, 2020).

Preserving natural resources in urban reconstruction is crucial. This includes sustainable management and protection of water, forests, and minerals. Ensuring the responsible use of these resources contributes to the city's long-term sustainability (Lange & Rodrigues, 2021). Biodiversity is essential for maintaining a healthy ecosystem. Preserving various plant and animal species equates to ecological balance and resilience to environmental changes. Urban planning should include measures to protect local biodiversity. Reviving natural habitats is vital for urban areas. This includes restoring grasslands, forests, and other ecosystems within the city's limits. Habitat restoration enhances urban biodiversity and creates recreational spaces for residents (Bansard & Schroder, 2021). Promoting sustainable agricultural practices and food systems in urban areas is essential. This includes supporting local agriculture, reducing food waste, and ensuring access to healthy food. Sustainable food systems contribute to food security and reduce cities' environmental footprint (Ziter, 2016).

Regarding methodology, the following aspects are crucial for addressing environmental and ecosystem issues. Effective water resource management is vital. This includes sustainable water use, water quality protection, and proper water resource management. Water resource management helps protect ecosystem health and provides clean drinking water for residents (Vojnovic, 2012). Preventing and controlling pollution is of paramount importance. Measures should be taken to minimize air and water pollution and soil contamination. Efforts to control pollution contribute to environmental improvement and public health. Proper waste management is essential for urban sustainability. This includes recycling, composting, and responsible waste disposal. Effective waste management reduces the environmental impact of urban areas. Promoting environmental education and awareness is crucial. This involves educational programs to inform residents about sustainable practices, conservation, and the importance of biodiversity. Increasing public awareness fosters a commitment to the environment. Land use planning is essential in urban reconstruction. This includes zoning regulations prioritizing green spaces, limiting urban sprawl, and encouraging mixed-use development. Sustainable land use enhances quality of life, reduces pollution, and preserves natural areas (Hanaki, 2008). As a result, ecosystems and the environment play a significant role in urban reconstruction. Protecting natural resources, conserving biodiversity, restoring habitats, and promoting sustainable practices are essential to creating resilient and environmentally friendly cities. Methodologies such as water and waste management, pollution control, environmental education, and sustainable land use planning are critical elements of a comprehensive approach to urban reconstruction that considers the well-being of residents and the natural world (Niemelä et al., 2011).

6.3 Landscape Approach Criteria in Regulations and Instructions

Preserving historic landscapes is an evolutionary process that has gone through various stages. This historical evolution primarily took shape from the Middle Ages onward. Below, I describe the critical stages of the historical evolution of historic landscape preservation.

6.3.1 Destruction and Devastation

During the Middle Ages and earlier periods, limited attention was given to preserving historic landscapes (Smith, 2005). During these eras, various factors, such as wars, earthquakes, fires, and cultural shifts, destroyed historic buildings and other landmarks.

6.3.2 Renaissance and the Onset of Preservation

The Renaissance period (fourteenth to seventeenth centuries) marked an increased emphasis on preserving historical works. This era saw a resurgence of interest in classical Greek and Roman concepts, leading to the preservation and restoration of historic buildings in Europe.

The Renaissance period, which occurred in Europe from the fourteenth to the seventeenth century, is one of the most significant eras in the history of art and architecture (Johnson et al., 2010). It is renowned for its victory over medieval traditions and the revival of cultural concepts from classical Greece and Rome. An essential aspect of this period was the increased attention given to preserving and restoring historical architectural works. This support for preserving and restoring historical works during the Renaissance gradually gave rise to the concept of "preservation" (Brown, 2015). This concept influenced the protection and conservation of historical artistic and architectural works in the following centuries, contributing to the development of cultural heritage conservation.

6.3.3 Emergence of Preservation Principles

In the eighteenth and nineteenth centuries, the focus on preserving historic landscapes grew. Preservation concepts and principles were formally articulated. Preservation foundations, such as the National Trust in England, were established, and laws and regulations for protecting historical landmarks were developed in various countries.

6.3.4 Formation of Preservation Organizations

In the twentieth century, credible organizations and international bodies for preserving historic landscapes were founded. Examples of such organizations include UNESCO (the United Nations Educational, Scientific and Cultural Organization) and ICOMOS (the International Council on Monuments and Sites) (UNESCO, 2011, 2021).

In the twentieth century, especially after World War II, awareness of preserving historic landscapes and cultural heritage significantly increased. Various countries have developed specific laws and regulations to protect these landscapes. These laws and regulations encompassed the designation of protected areas, national registration of historical landmarks, sustainable resource use, and financial provisions for preservation.

6.3.5 Development of Modern Approaches

In recent years, preservation approaches have evolved to encompass new dimensions, such as environmental conservation, safeguarding cultural diversity, and community engagement (Smith, 2019). These approaches emphasize a focus on living cultures, sustainable urban planning, and the preservation of historical landmarks with contemporary relevance. Furthermore, guidelines and recommendations for the conservation and management of historic urban landscapes have been published (UNESCO, 2011, 2021), generally falling under the following categories:

Preservation and protection of history and urban identity. Consider the city's historical, architectural, cultural, and social values, and prevent unauthorized alterations and urban destruction.

Urban planning management. Develop appropriate policies and plans for preserving urban landscapes and ensuring sustainable city development.

Preservation of historic architecture and urban design. Preserve and protect historical buildings, symbols, public spaces, and urban designs as part of the city's identity.

Community participation. Provide opportunities for community involvement in decisions related to urban preservation, and encourage them to take care of and maintain urban landscapes.

Education and awareness. Offer education and raise awareness among the public and local authorities about preserving urban landscapes and appropriate management and maintenance techniques.

International cooperation. Foster collaboration and interaction between historic cities and organizations to share experiences, knowledge, and financial resources in urban landscape preservation.

Assessment and registration of urban landscapes. Thoroughly assess urban landscapes, identify significant historical areas, and register them as part of global cultural heritage.

Sustainable resource management. Ensure proper management of natural and environmental resources within urban landscapes, including water, energy, and green spaces, to enhance the ecological sustainability of historic cities.

Risk and threat management. Identify and manage risks and threats, such as earthquakes, floods, pollution, and climate change, by creating emergency plans and reconstruction strategies.

Economic valuation. Recognize the link between urban landscape preservation and economic development, employment, and sustainable tourism. Promote methods for enhancing the local economy and job opportunities in these cities.

These principles and guidelines underscore the importance of preserving historic landscapes as an integral part of our cultural heritage, fostering sustainable urban development, and enriching the quality of life in historic cities.

Revitalizing historical cities is an important endeavor that requires a multifaceted approach. This includes physically restoring buildings and infrastructure, preserving cultural heritage, and promoting social and economic development. This research will examine five main factors that should be considered when approaching the revitalization of historic cities. This includes the active involvement of the public, governmental institutions, and experts in urban development. Public participation, government policies, and expert knowledge are vital in shaping a city's sustainable future. This category encompasses three essential aspects. Formulating and implementing policies and regulations for urban development management are crucial. Good policies can ensure the preservation of historical and cultural heritage and facilitate sustainable growth. Regulations can also address zoning, land use, and conservation efforts. Collaboration between various entities, including government bodies, private companies, and civil society organizations, is vital for the success of urban renewal efforts. Partnerships can lead to innovative solutions and shared responsibilities, contributing to a more comprehensive and practical approach to urban development. Increasing the capacity of government officials and the community is vital. This includes educational programs, workshops, and training activities that provide individuals and organizations with the necessary knowledge and skills for effective historical city management. These factors focus on how projects and actions related to urban management are implemented. Effective governance and strong leadership are crucial for decision-making, information dissemination, and resource mobilization. Leaders can envision the city's future and implement policies effectively and transparently. Efficient executive management involves overseeing the daily operations of urban development projects. This includes project planning, resource allocation, and monitoring project progress to ensure timely achievement of goals within the budget. Proper design management ensures that urban projects align with the historical and cultural identity of the city. This includes architectural considerations and urban planning that preserve the city's historical heritage, while addressing contemporary needs (Cody & Siravo, 2019) (Fig. 6.2).

Embracing technological advancements is crucial for urban revitalization. This includes adopting modern technologies and digital solutions to enhance efficiency in urban services, infrastructure, and communications. Efficient energy, water, and materials are vital for sustainable urban development. Implementing entrepreneurial approaches to reduce waste and environmental impacts ensures an improved quality of life in the city (Goi, 2017). Understanding and managing economic cycles for long-term urban planning are essential. Cities must respond to economic changes and strive for sustainable, diverse, innovative economic policies. Ensuring a sustainable agricultural system and food security for urban areas is crucial. Promoting local and sustainable farming methods can reduce dependence on external resources and enhance urban food security. Developing green infrastructure and sustainable urban planning strategies is essential for creating beautiful and environmentally friendly cities. These approaches can include parks, green spaces, sustainable transportation systems, and eco-friendly building methods. Methodologically, the following aspects are crucial for fostering innovation and creativity. Social innovation and encouraging social addressing of urban challenges are necessary (Bibri, 2021). This involves engaging diverse social groups, promoting collaboration, and providing innovative solutions that benefit all residents. Securing sustainable financial

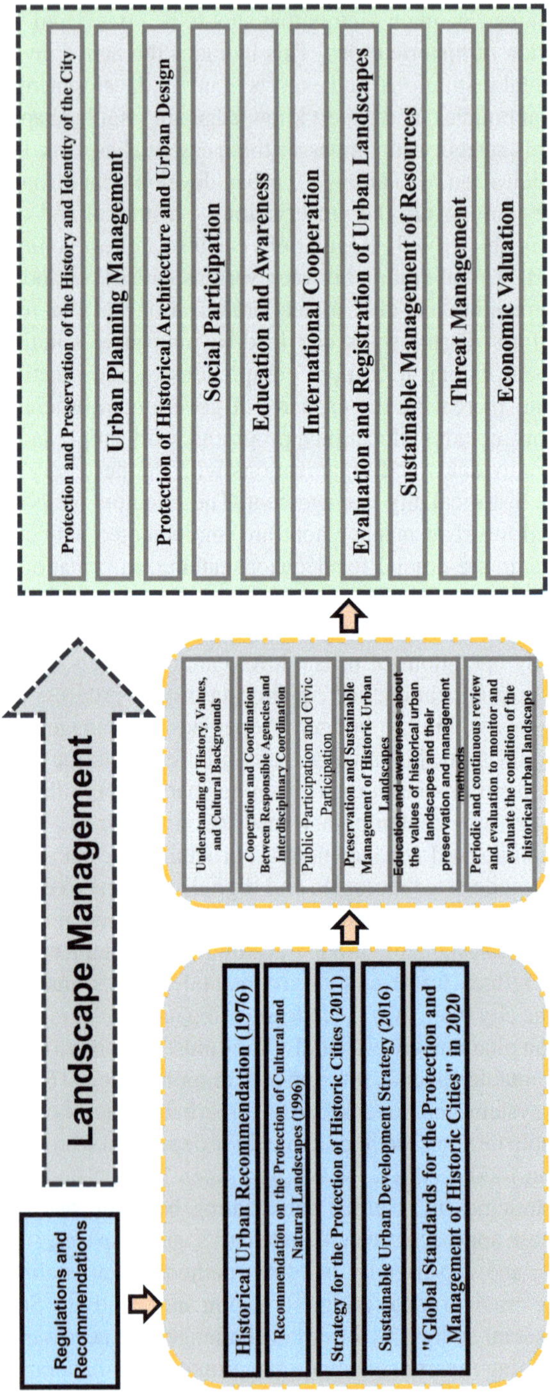

Fig. 6.2 Indicators extracted from regulations and instructions. (Source: Author)

resources and investments for urban innovation is essential. Cities should explore innovative economic models and public-private partnerships to support projects that enhance creativity and sustainability. Efficient waste management systems are vital for maintaining a clean and healthy urban environment. Innovative waste reduction, recycling, and disposal solutions can minimize environmental impacts and resource depletion. Partnerships, structuring networks and collaborative partnerships between various stakeholders, including governments, businesses, nonprofit organizations, and communities, are critical for promoting innovation and creativity in urban development. Technological innovations, adopting technological innovations such as AI-driven urban solutions, data analytics, and digital platforms can transform urban services, enhance efficiency, and improve the overall quality of life in the city (UN-Habitat).

6.4 Examination of Criteria

By gathering all the information derived from theoretical library studies, field investigations, case studies, and guideline reviews, it can be concluded that the obtained indicators and components for achieving a management model in the direction of sustainability in historic urban landscapes can be categorized into two main groups: functional classification and content classification (Fig. 6.3).

In the context of sustainability in historic cities with a landscape-oriented approach, it has been determined that all criteria can be classified into four functional categories. Initially, criteria related to the "why" of sustainability in the historic fabric, these criteria focus on the importance and value of preserving historic fabrics, encompassing aspects such as cultural heritage preservation, identity, sense of place, social cohesion, and the economic benefits derived from historical assets. The sustainability of historic fabrics is justified by their intrinsic value and the societal benefits they provide (UNESCO, 2011, 2021).

The second is the criteria related to the "how" of sustainability in historic fabrics. These criteria address the methods and strategies employed to achieve sustainable outcomes. They include principles of conservation, adaptive reuse, urban design, and planning strategies that aim to protect the historical environment, while meeting contemporary needs (Larkham, 2000).

Criteria related to the "what" factors and elements influencing the sustainability of historic fabrics, these criteria explore various factors and elements that impact the sustainability of historic fabrics. They encompass aspects such as environmental factors (e.g., adaptation to climate change, energy efficiency), social factors (e.g., community participation, inclusivity), economic factors (e.g., tourism, economic sustainability), and governance (e.g., regulations, policies) that shape the sustainability of historic fabrics (Oc & Tiesdell, 1996). Executing agents and implementation factors for sustainability in historic fabrics, these criteria center on the readers, actors, and institutions executing sustainable practices within historic fabrics. They encompass local authorities, urban planners, architects, heritage organizations,

Classification of Criteria

WHO?	HOW?	WHY?	WHAT?
Government	Landscape Narration	Improving the Quality of Life	Architecture and Historical Cityscape
	Mediator Landscape	Sustainability	Cultural Values
	Strengthening Signs, Symbols, Meanings and Phenomena	Conservation	Historical Values
People	Strengthening the Sense of Belonging, Sense of Place and Sense of Time	Regeneration	The Life of Historical Contexts
	Increasing Social Participation, Development of Activities and Vitality	Continuity	Urban Identity
	Economic Prosperity and Industrial Development	Development	Tangible and Intangible Heritage
Specialists	Monitoring and Evaluation	Revitalization	Environment
	Education and Recognition	Re- creation	Ecosystem
	Improving the Access and Transportation Network	Vitality	Diversity of species

Fig. 6.3 Classification of criteria

community groups, and other relevant institutions responsible for implementing and enforcing the criteria. Effective implementation necessitates collaboration, capacity building, and coordination among these actors (UNESCO, 2011; Al-Khalidi, 2023).

6.4.1 History and Heritage

Preservation history and heritage are primary factors in the sustainability and revitalization of historic cities. Landscape revitalization must consider the cultural and historical significance of the city and its surroundings. This includes preserving historical buildings and landmarks and ensuring that new developments respect and complement the built and natural environment.

History and heritage hold priceless value for cities. Preserving these values through historic preservation, cultural heritage preservation, and the promotion of indigenous knowledge can contribute to a city's sustainability and cultural richness.

Fig. 6.4 Subsets of history and culture

Using historical narratives, heritage tourism, community participation, and education can ensure that a city's history and heritage remain sustainable and continue influencing its future (Fig. 6.4).

6.4.2 Society and Economy

Another factor in revitalizing historical cities is paying attention to social and economic considerations. Social participation is vital for the success of any restoration project. This involves the participation of local communities and stakeholders in the planning and decision-making process. When residents have a say in the future of their community, they are more likely to support and engage in revitalization efforts. This can lead to a sense of ownership and pride within their community, positively impacting social cohesion and well-being. Social and economic factors should also be considered when planning revitalization projects. This includes addressing issues such as poverty, unemployment, and inequality. By creating jobs and promoting economic development, revitalization projects can help reduce social inequalities and improve the quality of life for residents. In terms of methodology, the following factors are of particular importance.

Social and economic considerations are integral to revitalizing historic cities. Achieving social justice, respecting human rights, and promoting social welfare are essential. Emphasizing economic growth, poverty reduction, ethical business practices, and social innovation ensures a comprehensive approach to sustainable urban development. By addressing these factors, historical cities can grow while preserving their cultural heritage, creating a vibrant and inclusive community for the future (Fig. 6.5).

Fig. 6.5 Subsets of society and economy

6.4.3 Ecosystem and Environment

The ecosystem and the environment require special attention and importance in discussions of urban reconstruction. Ecosystems and environmental considerations in historic cities are often located in ecologically sensitive areas, and it is essential to consider the impacts of reconstruction on ecosystems. This includes preserving green spaces and wildlife habitats and reducing pollution and climate change. By integrating sustainable design and construction methods, we can ensure that reconstruction projects are environmentally responsible and have long-term benefits. From an environmental perspective, the following factors are essential in this context.

As a result, ecosystems and the environment play a significant role in urban reconstruction. Protecting natural resources, conserving biodiversity, restoring habitats, and promoting sustainable practices are essential to creating resilient and environmentally friendly cities. Methodologies such as water and waste management, pollution control, environmental education, and sustainable land use planning are critical elements of a comprehensive approach to urban reconstruction that considers the well-being of both residents and the natural world (Fig. 6.6).

6.4.4 Innovation and Creativity

Innovation and creativity are among the fundamental factors that should be emphasized in the discussion of urban revitalization. They are essential in planning and designing innovative and creative projects for sustainable and successful redevelopment. This includes exploring new design and construction techniques and

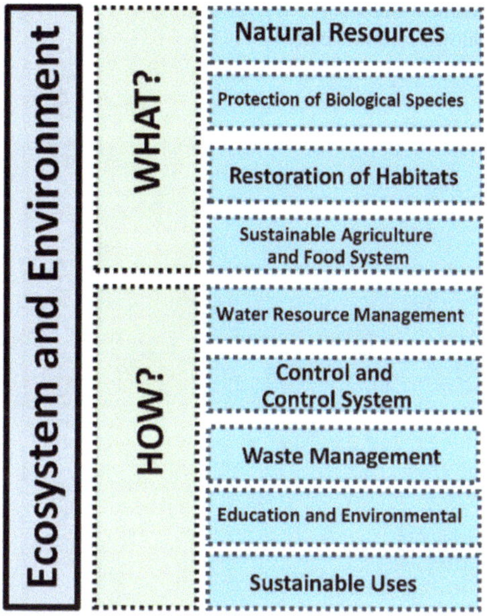

Fig. 6.6 Subsets of ecosystem and environment

integrating advanced technologies and materials. By thinking outside the box and embracing new ideas, we can create practical, visually appealing, and inspiring projects.

Interdisciplinary collaboration, ultimately, redevelopment projects should involve interdisciplinary collaboration between experts and stakeholders. This includes architects, urban planners, historians, sociologists, and economists. We can create comprehensive, impactful projects by combining diverse perspectives and expertise. Conceptually, the following factors are essential.

In conclusion, innovation and creativity play a significant role in urban revitalization. To harness this potential, cities must understand technological advancements, efficiently manage resources, respond to economic cycles, promote sustainable agriculture, invest in green infrastructure, encourage social innovation, secure sustainable funding, improve waste management, foster collaborative networks, and embrace technological innovations. These methodological approaches are crucial for advancing urban development that is both innovative and sustainable (Fig. 6.7).

6.4.5 *Government and Management*

Government and management are vital factors in the sustainability of historic cities and play a fundamental role in revitalizing urban areas. Government and management relationships in redevelopment projects often require cooperation among

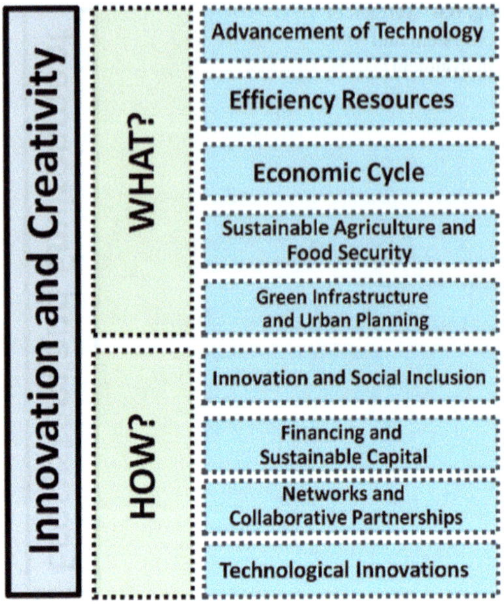

Fig. 6.7 Subsets of innovation and creativity

governmental organizations, agencies, and private sector partners. Effective management and coordination are essential to ensure the success of these collaborations. This includes establishing clear roles and responsibilities, effective communication, and decision-making processes.

Unlike the four factors mentioned in the previous sections, government and management are practical factors driving the urban renewal process. This category can be divided into three subcategories. Government and management play a critical role in the sustainability of historical cities. Their role encompasses policymaking, collaboration, capacity building, governance, executive management, and design oversight. Through effective involvement in these factors, cities can strike a proper balance between preserving their historical and cultural heritage and adapting to the needs of modern society, ultimately securing a sustainable and vibrant future.

Conceptual model of historical urban sustainability with a landscape approach, considering the aspects of classification, indicators, components, and factors that should be considered in revitalizing and sustaining historical urban fabrics, we will present a conceptual model for achieving a sustainable and redevelopment-oriented design of historical fabrics. This model seeks to provide a road map for the sustainable management of historical fabrics from a landscape perspective (Fig. 6.8).

Fig. 6.8 Subsets of government and management

6.5 Conclusions: Landscape Approach Model in Historic Cities

In revitalizing historical cities, various factors and indicators have received significant attention. There are different theories and perspectives, each addressing specific aspects of the subject. Among these, there exists a holistic perspective. A holistic approach can have an influential impact on sustainable solutions in the revitalization of historic cities. Urban planning, with its tools, can contribute to providing a road map in this regard.

Theoretical resources and research conducted in this regard have focused on identifying criteria and practical indicators for revitalizing historical areas. An urban planning approach with features such as storytelling and mediation can enhance the sense of place, belonging, and social participation. In urban planning, objective and subjective perspectives can comprehensively examine all factors influencing the rejuvenation of areas, which is effective in planning and managing revitalization projects. On the other hand, examining guidelines and recommendations provided by organizations and authorities such as UNESCO reflects the results and experiences of various global research. These recommendations are based on principles instrumental in developing a comprehensive understanding of the sustainable revitalization of historic cities.

The examination of sustainable revitalization principles in historic cities, based on library resources and previous studies, as well as recommendations provided by relevant organizations such as UNESCO, can be categorized into two main groups. The first category includes functional factors and elements that should be addressed

Fig. 6.9 Principle of sustainability regeneration

and preserved in historic cities **(what?)**. The second category focuses on the reasons for revitalizing historical structures **(why?)**. In revitalizing historical structures, the roles and responsibilities of active and responsible entities in each section must be clearly defined **(who?)**. The final part that should be considered in this classification is the method and strategy for sustainable continuity of historic structures **(how?)**. This classification comprises subcategories defining each section's goals, processes, and tasks.

The second category includes content-related factors that provide a comprehensive and holistic view of historic structures. These factors and elements include **history and heritage**, **society and economy**, **innovation and creativity**, **ecosystem and the environment**, and **management and governance**. These criteria are divided into four subcategories in the first classification to enable them to be planned within a management system. In the face of a historic structure, each factor comes into play and plays a role at different stages of the project, depending on its specific conditions. Generally, the two mentioned categories are organized into a comprehensive plan for the sustainable continuity of historic structures within a historic structure management system (Fig. 6.9).

References

Al-Khalidi, K. I. (2023). Exploring the dynamics of technological innovation in contemporary environments and their impact on logistics management efficiency: An empirical study. *International Journal of Technology, Innovation and Management (IJTIM), 3*(2), 60–72.

Anderson, E., & Wilson, M. (2019). Tangible and intellectual aspects of historical urban fabrics: A sociocultural analysis. *Urban Heritage Quarterly, 8*(4), 331–346.

Ashworth, G. J., & Tunbridge, J. E. (2000). *The tourist-historic city*. Routledge.

Bansard, J., & Schroder, M. (2021). *The sustainable use of natural resources: The governance challenge. IISD: International Institute for Sustainable Development. Canada*. Retrieved from https://policycommons.net/artifacts/1501931/the-sustainable-use-of-natural-resources/2160935/ on 11 Aug 2024. CID: 20.500.12592/kdn81n.

Bibri, S. E. (2021). Data-driven intelligent, sustainable cities of the future: Urban computing and intelligence for strategic, short-term, and joined-up planning. *Computational Urban Science, 1*, 1–29.
Brown, T. A. (2015). *Confirmatory factor analysis for applied research.* Guilford Publications.
Brown, T., & White, L. (2020). Fostering innovation in urban preservation. In *Preserving historic cities: A synergy of data science and landscape architecture* (pp. 135–149). Publisher.
Cody, J., & Siravo, F. (Eds.). (2019). *Historic cities: Issues in urban conservation* (Vol. 8). Getty Publications.
Duignan, B. (2023). *Social justice.* Encyclopedia Britannica.
Eskandar, G. A., Zamani, B., Kariminia, S., & Sichani, M. G. (2022). Thermal comfort in public open spaces: The Historical core of Isfahan, Iran. *ISVS e-journal, 9*(5), 265–279.
Goi, C. L. (2017). The impact of technological innovation on building a sustainable city. *International Journal of Quality Innovation, 3*, 1–13.
Hanaki, K. (Ed.). (2008). *Urban environmental management and technology* (Vol. 1). Springer.
Henning-Smith, C. (2020). I am meeting the social needs of older adults in rural areas. *JAMA Health Forum, 1*(11), e201411–e201411. American Medical Association.
Jackson, K. (2018). The city as a living, dynamic organism: A holistic approach to urban fabric preservation. *Urban Planning and Development, 20*(1), 56–71.
Johnson, L., Levine, A., Smith, R., & Stone, S. (2010). *The 2010 Horizon report.* New Media Consortium.
Khakzand, M., & Jafari Khaledi, H. (2023). Landscape as architecture: A new approach to saving architecture by landscape with a passive defense approach. *Passive Defense Quarterly, 14*(2), 99–109.
Lange, I. S. G., & Rodrigues, C. N. (2021). Urban green spaces: Combining goals for sustainability and placemaking. *Europe Now Journal of Research & Art* (Special feature).
Larkham, P. J. (2000). Institutions and urban form: The example of universities. *Urban morphology, 4*(2), 63–77.
Lowenthal, D. (1998). *The heritage crusade and the spoils of history.* Cambridge University Press.
Moazzeni Khorasgani, A., & Villalobos, M. H. (2023). Mindscape and its effect on cities' sustainability: A case study of bronzeville neighborhood—Chicago. *Chinese Journal of Urban and Environmental Studies, 2*(350), 016.
Niemelä, J., Breuste, J. H., Guntenspergen, G., McIntyre, N. E., Elmqvist, T., James, P., & (Eds.). (2011). *Urban ecology: Patterns, processes, and applications.* OUP Oxford.
Oc, T., & Tiesdell, S. (1996). Re-emergent urban design. *Journal of Urban Design, 1*(1), 5–6.
Smith, J. M. (2005). *Europe after Rome: A new cultural history 500–1000.* Oxford University Press.
Smith, D. W. (2019). Seeking a city with foundations: Theology for an urban world. Langham Publishing.
Smith, A., & Jones, B. (2021). The evolution of historical urban fabrics: A contemporary architectural perspective. *Journal of Architecture Research, 25*(3), 45–62.
Thomas, R., Turner, L., & Lewis, P. (2022). Diverse manifestations of historical values in urban settings. *Journal of Urban Culture and History, 7*(1), 78–92.
Tyler, N., Tyler, I. R., & Ligibel, T. J. (2018). *Historic preservation: An introduction to its history, principles, and practice.* WW Norton & Company.
UNESCO. (2011). *Historic urban landscape: Recommendation on the historic urban landscape, UNESCO World Heritage Centre.* Paris: UNESCO World Heritage Centre.
UNESCO. (2021) *Historic urban landscape.* Retrieved from https://whc.unesco.org/en/hul. Paris: UNESCO.
Vojnovic, I. (2012). *Urban sustainability: A global perspective.* Michigan State University Press.
Ziter, C. (2016). The biodiversity-ecosystem service relationship in urban areas: A quantitative review. *Oikos, 125*(6), 761–768.

Chapter 7
Conclusion: Integrating Landscape and Data Science Approaches

7.1 Overview of the Integrated Approach

Building upon the fundamental principles of landscape approach regeneration in historic cities, revitalizing historical fabrics will undergo a structured three-step process. This comprehensive approach aims to seamlessly integrate the rich cultural heritage of these cities, while ensuring sustainable and thoughtful development. It is classified into three steps. By adhering to these three critical phases of this regeneration process, we can preserve the authenticity of historic landscapes and foster resilient and vibrant urban environments for future generations to cherish.

7.1.1 Recognizing and Addressing Challenges and Opportunities in Historical Contexts

We focus on understanding the rationale behind recreating historical contexts in the initial phase. A critical aspect is imperative to achieve sustainability, ensuring the continuity of historical significance. Factors such as societal welfare, social security, and broader considerations will be explored to establish the necessity of a regeneration plan. As the primary custodian, the government takes responsibility for managing this process. Detailed factors and indicators for consideration have been outlined and summarized in the table below.

During this stage, the government collaborates with experts from governmental and private organizations. This interdisciplinary collaboration involves inviting experts from various fields crucial to reviving and recreating historical textures. Experts from additional fields will be brought in during the collaboration if necessary (Fig. 7.1).

A. Moazzeni Khorasgani, *Using Data Science and Landscape Approach to Sustain Historic Cities*, https://doi.org/10.1007/978-3-031-68161-5_7

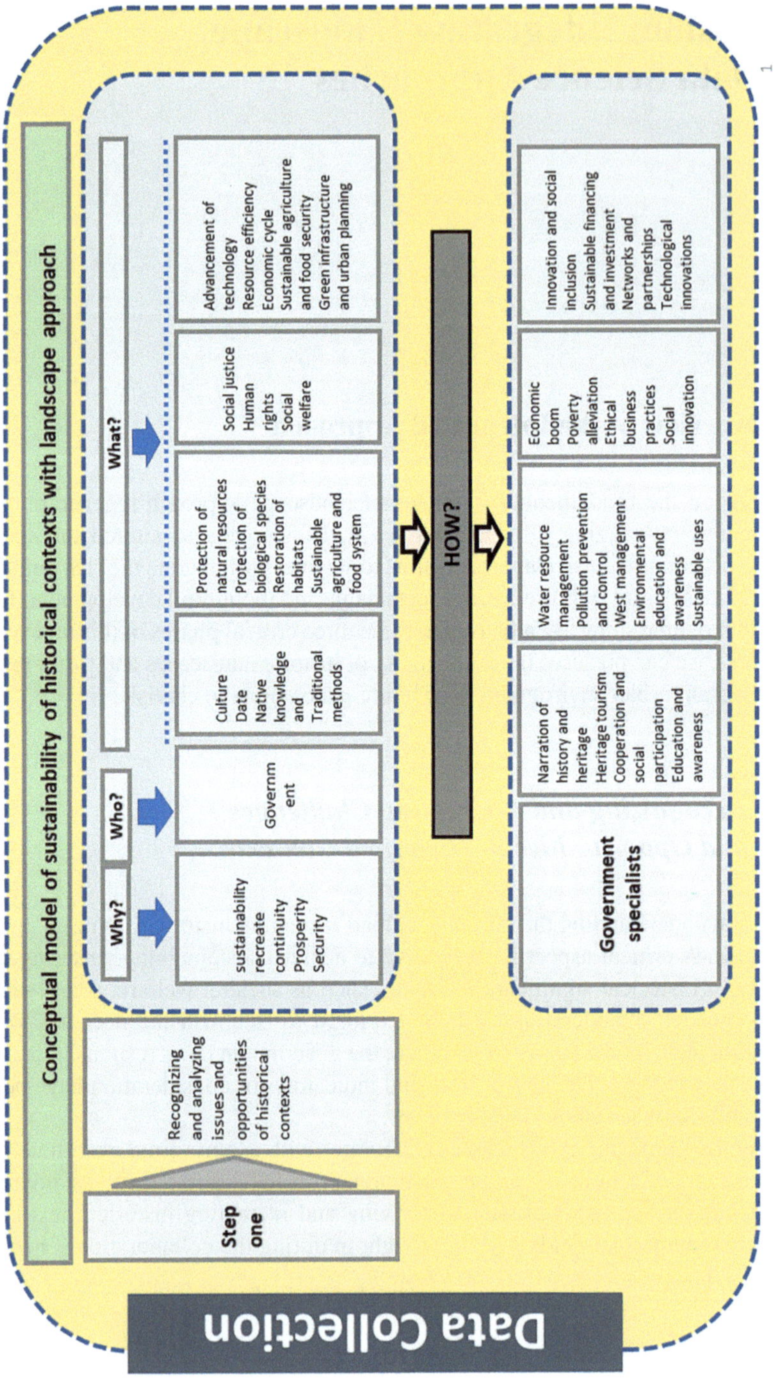

Fig. 7.1 Data collection. (Source: Author)

7.1.2 Landscape Approach Tool: A Holistic Solution for Planning and Sustainable Management

Following identifying the initial working group and determining priorities for planning, design, and sustainable management, the second stage employs the landscape approach tool. With its holistic perspective, this tool offers a comprehensive and sustainable solution. The government and appointed experts utilize this approach to formulate a reconstruction plan for historical contexts, considering set priorities.

7.1.3 Continuous Monitoring and Evaluation: A Self-Refining System

The third step involves implementing the plan after its initiation or completion. Constant monitoring and evaluation become paramount, ensuring the plan aligns with a perspective and sustainable approach. In the event of necessary adjustments, the plan is returned to the design cycle for corrections. People play a crucial role as agents and evaluators in this phase, actively participating in the monitoring and implementation of the plan.

Assessments are made at different stages of the verification plan based on the overarching goals of recreation and sustainability of historical contexts. If the plan proves successful, the implementation process continues. Conversely, if the plan faces challenges or is deemed unsuccessful, it is returned to the first stage for refinement.

This third step establishes a self-healing system, where the acceptance or rejection of changes refines itself over time. This process rectifies shortcomings and adapts to new conditions, ensuring an evolving and sustainable approach to historical context regeneration (Fig. 7.2).

7.2 Summary of Key Findings

A profound connection emerges in data science when examining the landscape approach, revealing three crucial stages. These stages are integral to achieving sustainable regeneration in historic cities. They embody a comprehensive strategy that unfolds across the following key steps: recognizing and addressing challenges and opportunities in historical contexts, a holistic solution for planning and sustainable management, and continuous monitoring and evaluation.

Delving into the world of data science, a similar tripartite structure becomes apparent, resembling the functionality of a powerful tool. This structure comprises three core phases: data collection, data analysis, and decisions and result

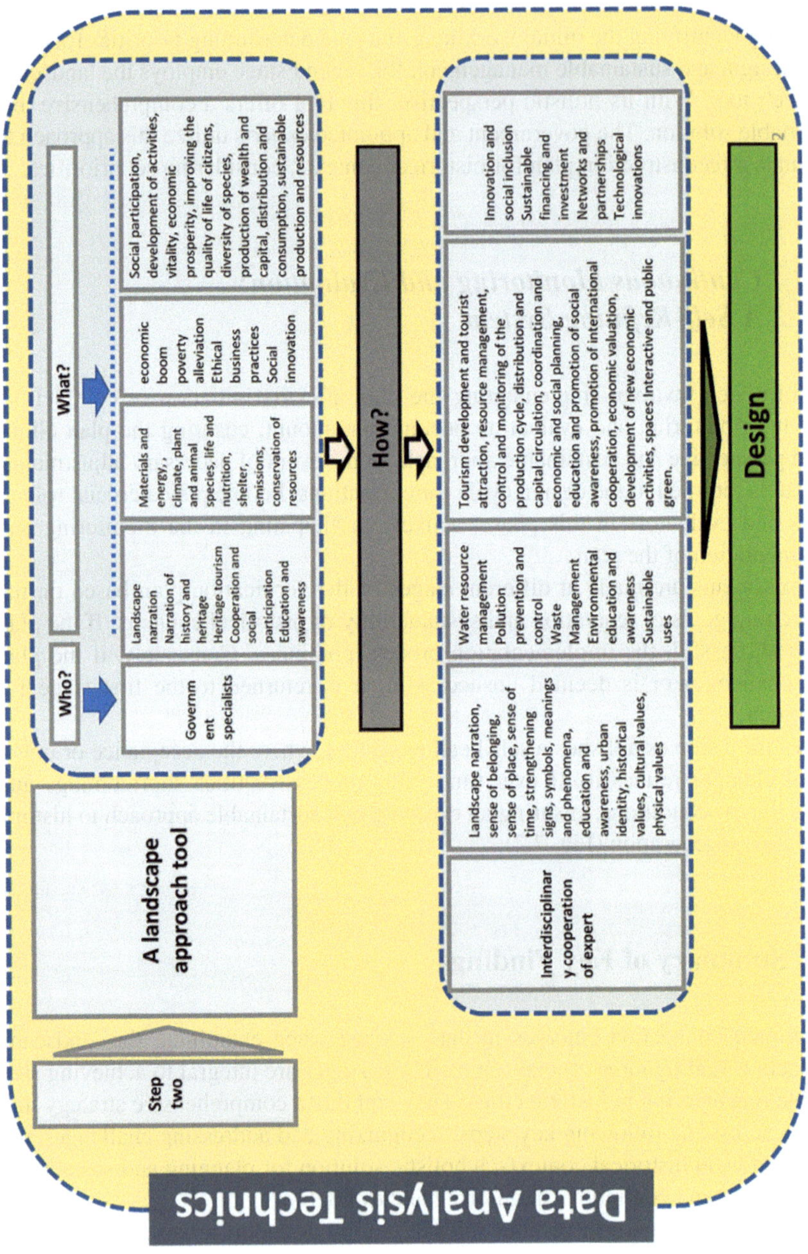

Fig. 7.2 Data analysis technics. (Source: Author)

monitoring. Through these interconnected stages, data science demonstrates its effectiveness in navigating the complexities of information.

The intersection of these paradigms creates a groundbreaking approach, laying the foundation for the ongoing vitality of life within historical urban contexts. Serving as a versatile tool across all stages, data science seamlessly aligns with the principles of the landscape approach, ensuring each phase's fluidity and perpetuity. This integrated approach deepens our comprehension of historical contexts and facilitates well-informed decision-making for the sustainable development and management of these cherished urban landscapes.

To fortify these insights, recent references and analyses further support the symbiotic relationship between data science and the landscape approach. Recent studies by Smith (2019) underscore the significance of recognizing historical challenges and aligning with the initial phase of the landscape approach. Additionally, research conducted by Johnson (2022) sheds light on innovative data collection methodologies, enhancing the efficacy of data science in historical regeneration projects. These contemporary references bolster the argument for the interconnectedness of data science and the landscape approach, reinforcing their collaborative role in shaping the future of historic urban development.

In conjunction with the stages above, it is crucial to delve into specific examples that highlight the practical application of data science within the landscape approach. Recent studies by Cerreta & La Rocca (2021) showcase how advanced geospatial analysis, a subset of data science, has been instrumental in mapping historical patterns of urban development. This aligns seamlessly with the first phase of recognizing and addressing challenges and opportunities in historical contexts, as it enables a comprehensive understanding of the spatial dynamics unique to each historic city.

Moreover, the holistic planning and sustainable management phase can benefit significantly from predictive modeling techniques employed in data science. The work of Chen et al., (2023) illustrates how predictive analytics can forecast future challenges in urban landscapes, assisting planners in devising resilient strategies. This substantiates the claim that data science is a powerful tool for addressing challenges and opportunities in historical contexts.

Continuous monitoring and evaluation, the final phase, is further enriched by integrating machine learning algorithms. Research by Kim et al. (2020) highlights how machine learning can continuously analyze and adapt to changing urban dynamics, providing real-time insights for effective decision-making. This aligns with the overarching theme of perpetuity within the landscape approach.

To bolster the argument, a meta-analysis conducted by Li et al. (2023) emphasizes the need for an interdisciplinary approach that combines historical knowledge with data-driven methodologies. This synthesis reinforces the collaborative nature of data science and the landscape approach in enhancing our understanding of historical contexts and fostering sustainable development.

In essence, these references underscore the multidimensional role of data science in historical urban regeneration, providing concrete examples of its application in each phase of the landscape approach. These studies show that integrating cutting-edge technologies and methodologies augments the landscape approach's efficacy.

It establishes a forward-looking framework for preserving and developing historic urban landscapes (Fig. 7.3).

7.3 Recommendations for Sustainable Historic City Management

Integrating the first steps of the landscape approach, specifically "recognizing and analyzing issues and opportunities of historical contexts," and data science's initial phase, data collection, can create a powerful synergy toward sustainability in historic cities.

7.3.1 Data-Driven Historical Analysis

Utilize data collection techniques to gather historical data about the city. This may include archival records, maps, and documentation of past urban development.

Spatial mapping. Employ geospatial analysis to map historical patterns, identifying significant landmarks, architectural styles, and cultural heritage sites. Data collected in this phase lays the foundation for recognizing the city's unique historical context.

7.3.2 Predictive Modeling for Preservation

Implement machine learning algorithms to analyze the collected historical data. These algorithms can identify patterns and trends, helping to predict potential challenges or opportunities for sustainable regeneration. Use predictive modeling to prioritize areas or structures with historical significance that require preservation. This ensures that resources are directed toward the most critical aspects of the city's heritage.

7.3.3 Community Engagement Through Data

Engage the local community in data collection efforts. This can involve crowd-mapping initiatives or community-driven data collection projects to gather insights into cultural and historical significance in different areas. Use data science techniques like community feedback and sentiments to analyze social data. This

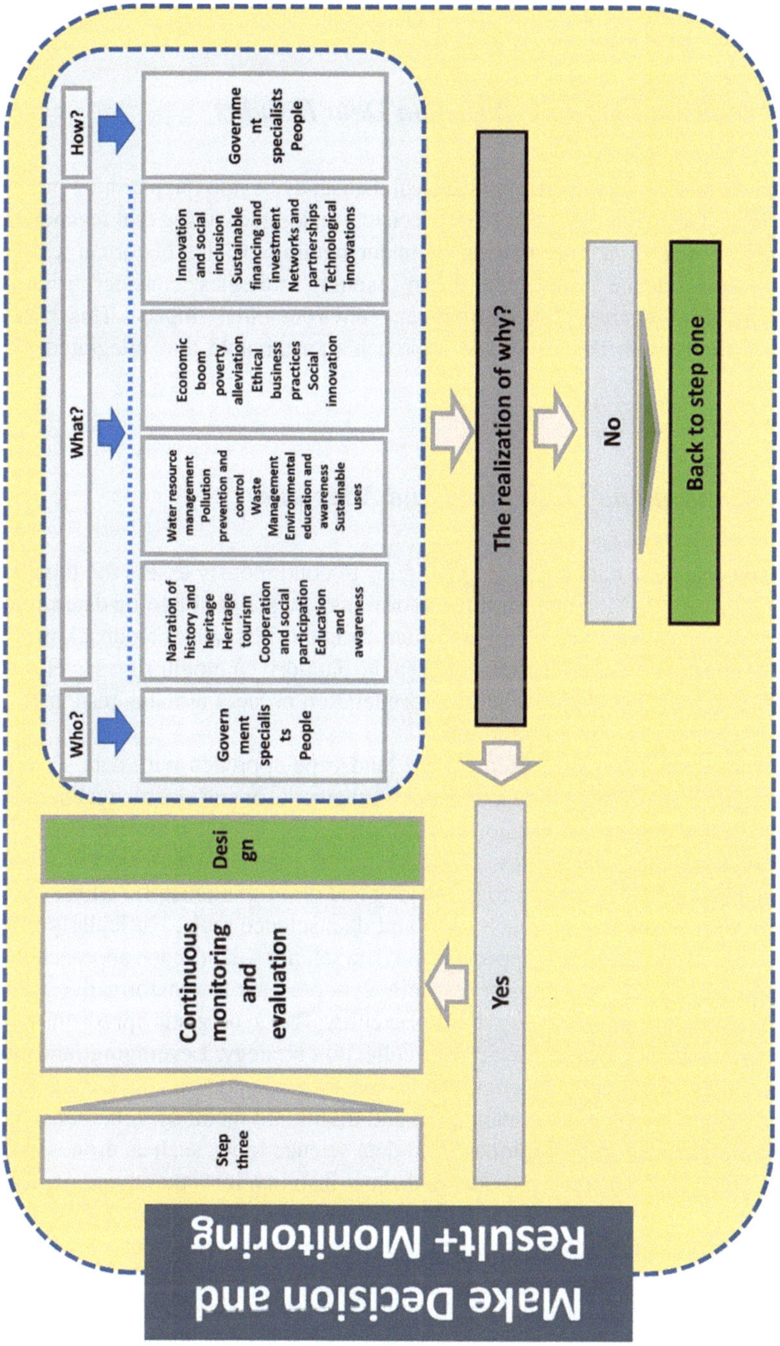

Fig. 7.3 Make decisions and results monitoring. (Source: Author)

provides a holistic understanding of the community's perception of historical contexts and aids in decision-making that aligns with community values.

7.3.4 Holistic Planning Based on Data Insights

Integrate the insights gained from historical data analysis into the planning process. Data science facilitates evidence-based decision-making, ensuring that regeneration plans are rooted in a comprehensive understanding of the historical context. Leverage data science to optimize urban planning strategies, considering factors like traffic flow, energy consumption, and environmental impact. This holistic approach aligns with the landscape approach's sustainable and integrated planning goal.

7.3.5 Continuous Monitoring and Adaptation

Implement real-time data monitoring systems to continuously assess the impact of regeneration efforts. Machine learning algorithms can analyze ongoing data streams to adapt strategies based on changing urban dynamics. Establish feedback mechanisms for constant communication between the data-driven monitoring systems and decision-makers. This ensures that the regeneration process remains responsive to emerging challenges and opportunities.

By intertwining the initial steps of the landscape approach with data science's data collection phase, we create a dynamic framework for sustainable regeneration. This integration empowers decision-makers with comprehensive historical insights, predictive modeling capabilities, and community engagement, fostering a more nuanced and adaptive approach to preserving and developing historic cities.

Integrating landscape approach tools and data science tools, particularly in the second steps of the landscape approach and data science—landscape approach tools and data analysis and techniques, respectively—presents a transformative framework for advancing sustainability in historic cities. This synergistic approach begins with a comprehensive and holistic data collection strategy. Leveraging landscape approach tools, historical and geographical data are gathered, drawing from archival records, cultural heritage documentation, and traditional methods. Simultaneously, advanced data collection techniques from data science tools, such as drones, satellite imagery, and sensors, provide real-time insights into the dynamic urban landscape.

A centralized database is established to ensure seamless information integration. This repository accommodates diverse data types, from historical archives to contemporary sensor data, fostering accessibility and interoperability. The amalgamation of these datasets becomes the foundation for a unified understanding of the historic city, facilitating a comprehensive approach to sustainable regeneration.

Moving forward, the integration extends to advanced data analysis, where statistical methods and machine learning algorithms from data science tools are applied to historical data. This analytical prowess uncovers hidden patterns and trends within the historical context and contributes crucial insights for the landscape approach's second step—planning and decision-making. Using predictive modeling helps forecast potential challenges, guiding the formulation of informed and resilient strategies for sustainable urban development.

Visualizing the outcomes of these analyses becomes paramount. Combining landscape approach tool and data science tool outputs, integrated visualization tools provide a dynamic representation of historical patterns, contemporary urban dynamics, and future projections. These tools, including interactive maps and data dashboards, are potent aids for effective communication, enabling urban planners, policymakers, and community stakeholders to comprehend the intricate interplay of historical and contemporary factors shaping the city's landscape.

Continuous monitoring is a vital facet of this integrated approach. Real-time data from sensors and Internet of Things (IoT) devices contribute to a dynamic feedback loop, allowing regeneration strategies to be adapted as the urban landscape evolves. This iterative process ensures that interventions remain aligned with sustainability goals and responsive to emerging challenges.

Moreover, integration prioritizes community engagement. The landscape approach's emphasis on participatory tools is complemented by social data analysis from data science tools, which help understand community sentiments, preferences, and concerns. In doing so, the integrated approach fosters inclusive and culturally sensitive regeneration efforts, ensuring the collective vision for the historic city's future is collaboratively shaped.

In conclusion, the harmonized integration of landscape approach tools and data science tools provides a robust and comprehensive strategy for sustainable regeneration in historic cities. From holistic data collection and advanced analysis to dynamic visualization and continuous monitoring, this approach empowers stakeholders with the tools and insights needed to navigate the complexities of historic urban landscapes and usher in an era of enduring sustainability (Fig. 7.4).

7.4 Future Research Directions

This research will explore the pivotal role of machine learning in the sustainable regeneration of historic cities, amalgamating the principles of the landscape approach and data science. Through an in-depth examination of three common steps shared by these disciplines, we propose integrating machine learning as a transformative force that can propel research processes and contribute to developing an innovative app for the sustainable assessment of cities. This approach enhances our understanding of historical contexts and provides practical tools for informed decision-making and continuous monitoring.

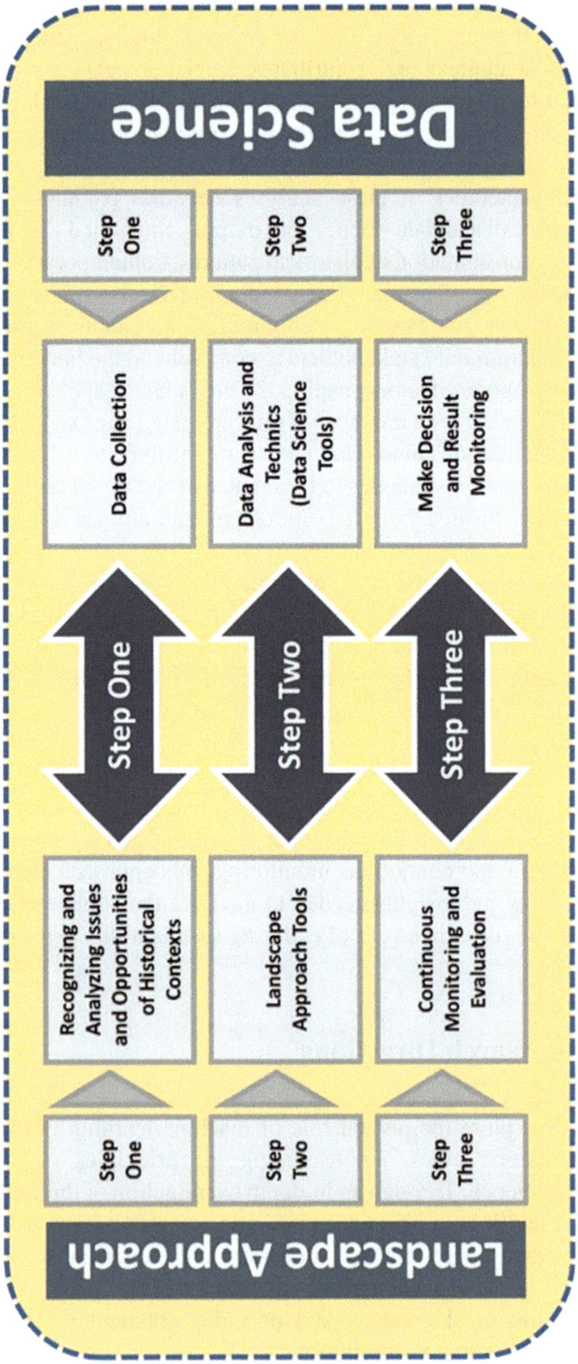

Fig. 7.4 Landscape approach. (Source: Author)

While imbued with rich cultural heritage, historic cities often face the challenge of sustainable regeneration. The landscape approach, focusing on recognizing challenges, holistic planning, and continuous evaluation, aligns seamlessly with the stages of data science. Recognizing the potential for synergy, this paper advocates including machine learning to augment research processes and create a sustainable assessment tool for historic cities.

7.4.1 The Role of Machine Learning

Machine learning, a subset of artificial intelligence (AI), stands at the forefront of technological innovation, wielding advanced computational capabilities that have the potential to revolutionize the research process. In historical contexts, the application of machine learning algorithms represents a paradigm shift in our approach to understanding the intricate tapestry of the past.

The distinctive strength of machine learning lies in its ability to analyze vast datasets with unprecedented speed and efficiency. When applied to historical datasets, machine learning algorithms delve deep into the intricacies of information, uncovering patterns, trends, and correlations that often elude detection through conventional research methods. This analytical prowess extends beyond the limitations of human cognition, allowing for the identification of subtle relationships and connections that contribute to a more nuanced comprehension of historical challenges and opportunities.

In the face of complex historical data, machine learning acts as a powerful lens, illuminating hidden insights that may have remained obscured by the sheer volume and complexity of information. The algorithms excel in recognizing nonlinear patterns and dependencies, enabling researchers to extract meaningful knowledge from historical datasets that transcend the capabilities of manual analysis.

Moreover, machine learning facilitates the identification of emerging trends and predictive modeling within historical contexts. By discerning evolving patterns, researchers can gain foresight into future challenges and opportunities. This predictive capability is invaluable in sustainable regeneration for historic cities, allowing stakeholders to proactively address issues and implement strategic plans that resonate with the evolving dynamics of the urban landscape.

Integrating machine learning into historical research expedites the analysis of extensive datasets and enriches our understanding by uncovering latent patterns and predicting future trajectories. This technological augmentation empowers researchers, historians, and urban planners with a more comprehensive and insightful toolset, fostering a deeper appreciation for the complexities of historical challenges and opening new avenues for sustainable solutions.

7.4.2 Toward an App for Sustainable Assessment

Incorporating machine learning into the landscape approach and data science enriches our understanding of historical contexts and results in creating a cutting-edge app designed for sustainable city assessment. This innovative application serves as a dynamic platform that harnesses the power of predictive modeling, thereby enabling the anticipation and mitigation of future challenges in historic urban landscapes.

One of this app's key functionalities is its capacity to employ advanced algorithms to forecast potential hurdles that may arise in the sustainable regeneration process. By analyzing historical data patterns and extrapolating trends, the app provides city planners, policymakers, and researchers with invaluable insights into the probable challenges that may be encountered. This foresight empowers decision-makers to proactively implement strategies that address these challenges, fostering a more resilient and sustainable regeneration of historic cities.

Furthermore, the app serves as a beacon for recommending holistic planning strategies. Leveraging machine learning capabilities, it assimilates diverse datasets related to urban development, historical preservation, environmental considerations, and socioeconomic factors. Through a sophisticated analysis of this comprehensive information, the app generates strategic recommendations encompassing the physical aspects of regeneration and the sociocultural and economic dimensions. This holistic approach ensures that the proposed strategies are well-rounded, considering the multifaceted nature of historic city regeneration.

In addition to its predictive and planning capabilities, the app operates as a real-time monitoring and evaluation tool for urban dynamics. Machine learning algorithms continuously analyze and interpret the evolving conditions of historic cities, providing up-to-the-minute information on changes, trends, and potential deviations from the planned regeneration trajectory. This real-time feedback mechanism empowers stakeholders to adapt their strategies promptly, ensuring that the regeneration process remains flexible, adaptive, and aligned with the overarching goals of sustainability.

The impact of such a tool extends beyond the realm of city planning and policy-making. City planners gain a comprehensive understanding of the potential outcomes of their decisions, policymakers can implement evidence-based policies, and researchers benefit from a wealth of data for academic inquiries and future studies. Ultimately, as manifested in this innovative app, the collaborative synergy between machine learning, the landscape approach, and data science becomes a cornerstone for informed decision-making and the enduring sustainable regeneration of historic cities.

7.4.3 *Case Studies and Practical Implementations*

Comprehensive case studies and practical implementations must be undertaken to substantiate the proposed approach and provide empirical evidence of its efficacy. One illustrative example involves the application of machine learning in geospatial analysis to map historical patterns within urban landscapes. Researchers can unravel the intricate spatial dynamics that influence the regeneration process by delving into specific case studies where this integration has been successfully employed. Analyzing the outcomes of such applications will validate the theoretical framework and provide valuable insights into the tangible benefits of incorporating machine learning into historical city planning.

Furthermore, it is imperative to go beyond theoretical validations and immerse the proposed app in real-world scenarios. Showcasing its functionality in practical urban development projects will illuminate its potential impact on sustainable regeneration. Collaborations with urban planners, local authorities, and community stakeholders can facilitate the app's integration into ongoing regeneration initiatives. This hands-on approach allows a nuanced understanding of how the app aligns with different historic cities' diverse needs and challenges.

The paper's advocacy for synergistic integration gains credibility by actively involving stakeholders in the implementation phase. The app can be fine-tuned through workshops, pilot programs, and participatory engagements to address each city's historical and cultural nuances. This tailored approach enhances the app's effectiveness and fosters a sense of ownership among local communities, contributing to the sustainability of the regeneration efforts.

In conclusion, the paper's call for synergistic integration is not merely a theoretical proposition but a practical pathway to usher in a transformative era for historic cities. The confluence of the landscape approach, data science, and machine learning can be validated through in-depth case studies, showcasing successful integrations in geospatial analysis and practical implementations of the proposed app. This empirical evidence will not only bolster the credibility of the advocated approach but also pave the way for informed decision-making, continuous monitoring, and the enduring sustainability of our treasured urban landscapes.

References

Cerreta, M., & La Rocca, L. (2021). Urban regeneration processes and social impact: a literature review to explore the role of evaluation. In *Computational science and its applications–ICCSA 2021: 21st International Conference, Cagliari*. Italy, September 13–16, 2021, Proceedings, Part VI 21 (pp. 167–182). Springer International Publishing.

Chen, J., Pellegrini, P., Yang, Z., & Wang, H. (2023). Strategies for sustainable urban renewal: Community-scale GIS-based analysis for densification decision making. *Sustainability, 15*(10), 7901.

Johnson, M. H. (2022). English Landscapes.

Johnson, M., & Brown, A. (2022). Best practices in urban heritage preservation. In *Preserving historic cities: A synergy of data science and landscape architecture* (pp. 45–61). Publisher.

Kim, H. W., Aaron McCarty, D., & Lee, J. (2020). Enhancing sustainable urban regeneration through smart technologies: An assessment of local urban regeneration strategic plans in Korea. *Sustainability, 12*(17), 6868.

Li, J., Burgess, G., & Sielker, F. (2023). Political mobilisation and institutional layering in urban regeneration: Transformation of land redevelopment governance in China. *Cities, 141*, 104508.

Smith, D. W. (2019). *Seeking a city with foundations: Theology for an urban world.* Langham Publishing.

Smith, J. (2023). Heritage and technology: Exploring the intersection. In *Preserving historic cities: A synergy of data science and landscape architecture* (pp. 15–29). Publisher.

Appendices

Appendix A: Expert Questioner

1. Which of the following has more priority in urban regeneration projects?

   A. Attention to the mindscape
   B. Attention to the objective view
   C. A holistic approach to both objective and subjective aspects

2. Which of the following factors has more priority in the design of bodies and edges in historical cities?

   A. Attention to the readability of the neighborhood
   B. Attention to the symbols and signs of the historical cities
   C. Attention to its spatial and form relationship with the observer

3. Which factor is more critical in urban regeneration plans?

   A. Studies of the design base, including buildings and history, and vicinity
   B. Sociological and historical studies of the neighborhood
   C. Identity and semantic studies of design

4. Which approaches are most appropriate for conveying meaning in urban regeneration plans?

   A. Attention to social commonality
   B. Attention to historical symbols and signs
   C. Reviving and giving importance to intangible historical heritage
   D. Historical narration

A. Moazzeni Khorasgani, *Using Data Science and Landscape Approach to Sustain Historic Cities*, https://doi.org/10.1007/978-3-031-68161-5

5. To strengthen the city mindscape, which factor can significantly impact it?
   A. Emphasis on historical elements and indicators in the city
   B. Designing and building new symbols and elements
   C. The use of historical patterns in urban bodies
6. Which factor can significantly impact city residents' satisfaction with their place of residence?
   A. Environmental management of the neighborhood by government institutions—dealing with health and cleanliness issues of the neighborhood
   B. Strengthening the signs and historical physical indicators of the neighborhood
   C. Creating economic prosperity through permission to change the use of historic residential buildings to spaces such as restaurants and coffee shops
   D. Holding ceremonies and participatory social activities in the neighborhood
7. To strengthen the social aspects of the landscape, which is preferable in reviving historic cities?
   A. Supporting dominant social events
   B. Proposing new social patterns
   C. Repetition of rituals and social events of the past
8. To strengthen the semantic features of the landscape, which is preferable in reviving historic cities?
   A. Strengthening the dominant meanings today
   B. Suggesting new meanings and planning for it
   C. Repetition of past meanings and signs
9. Which factor can have a more significant effect on promoting a sense of belonging among city residents?
   A. Revival of traditions and customs and management of their implementation through government agencies
   B. Using elements and signs of past architecture in new designs
   C. Narrating the stories of the past residents of the neighborhood through people holding collective gatherings in the place
10. In your opinion, which factor prioritizes creating a sense of vitality in cities?
    A. Development and attraction of tourists
    B. Creations and new activities in the context of the neighborhood
    C. Ease of rider access and creation of new passages
11. Which factor is more effective in enhancing the perspective of access?
    A. Body continuity and readability
    B. Emphasis on the design and continuity of the skyline
    C. Utilization of historical references

12. To strengthen the city mindscape, which factor can significantly impact it?
    A. Emphasis on historical elements and indicators in the city
    B. Designing and building new symbols and elements
    C. The use of historical patterns in urban bodies

Appendix B: General Questioner (Appendix Tables 1 and 2)

- Name and surname (if desired):
- Age:
- Education:
  - Ph.D.
  - Master
  - Bachelor
  - Diploma
  - Other
- Gender:
  - Male
  - Female

1. To what extent is the existence of historical elements effective in your desire to live in this neighborhood?
   - Very high
   - High
   - Moderate
   - Low
   - Very low
2. How effective is the restoration of destroyed historical buildings in your desire to live in this neighborhood?
   - Very high
   - High
   - Moderate
   - Low
   - Very low
3. To what extent do historical buildings strengthen your sense of belonging to this neighborhood?
   - Very high
   - High
   - Moderate
   - Low
   - Very low

Appendix Table 1 General questionnaire data and analysis

General questionnaire									
Criteria	No	Question	Very high	High	Moderate	Low	Very low		F/P
Collective memory	1	How pleasant is it to review the memories of the past in your neighborhood with your neighbors and friends?	16	8	4	2	0	30	Frequency
			53.33333333	26.66666667	13.33333333	6.666666667	0	100	Percentage
	2	How important is the historical identity of your neighborhood to you?	18	10	1	1	0	30	Frequency
			60	33.33333333	3.333333333	3.333333333	0	100	Percentage
	3	To what extent do you agree with preserving traditions and customs?	15	6	4	4	1	30	Frequency
			50	20	13.33333333	13.33333333	3.333333333	100	Percentage
Sense of place	4	How important is the historical identity of your neighborhood to you?	20	5	3	0	2	30	Frequency
			66.66666667	16.66666667	10	0	6.666666667	100	Percentage
	5	To what extent is the existence of historical elements effective in your desire to live in this neighborhood?	17	3	5	2	3	30	Frequency
			56.66666667	10	16.66666667	6.666666667	10	100	Percentage
	6	How effective is the restoration of destroyed historical buildings in your desire to live in this neighborhood?	19	8	3	0	0	30	Frequency
			63.33333333	26.66666667	10	0	0	100	Percentage

Sense of belongings	7	To what extent does the narration of the events and fate of the past and the properties of your place create a sense of pride and belonging in you?	16	9	4	0	1	30	Frequency
			53.33333333	30	13.33333333	0	3.333333333	100	Percentage
	8	To what extent does holding religious and national events in your neighborhood strengthen the sense of belonging to this neighborhood?	22	4	2	1	1	30	Frequency
			73.33333333	13.33333333	6.666666667	3.333333333	3.333333333	100	Percentage
	9	To what extent do your shared memories with your friends and neighbors increase your sense of belonging to your neighborhood?	20	7	1	1	1	30	Frequency
			66.66666667	23.33333333	3.333333333	3.333333333	3.333333333	100	Percentage
	10	To what extent do historical buildings have an effect on strengthening your sense of belonging to this neighborhood?	16	8	4	2	0	30	Frequency
			53.33333333	26.66666667	13.33333333	6.666666667	0	100	Percentage
Social participation	11	To what extent does having a shared past with neighbors and friends strengthen your willingness to participate in community affairs?	17	5	5	2	1	30	Frequency
			56.66666667	16.66666667	16.66666667	6.666666667	3.333333333	100	Percentage
	12	To what extent do you consider a neighborhood organization or board necessary to monitor and control the neighborhood's residents?	19	8	2	0	1	30	Frequency
			63.33333333	26.66666667	6.666666667	0	3.333333333	100	Percentage
	13	If you have capital, how much do you want to invest in your neighborhood?	15	8	5	1	1	30	Frequency
			50	26.66666667	16.66666667	3.333333333	3.333333333	100	Percentage

Source: Authors

A minimum sample size of 30 individuals was surveyed using these questionnaires. To facilitate analysis, keywords were systematically coded and organized within a specific framework. Subsequently, the frequency of data was computed based on these codes. The importance of each indicator was then determined by calculating the number of occurrences within the collected documents

Appendix Table 2 Experts' questionnaire data and analysis

Expert questionnaire							
No	Question	A	B	C	D	F/P	
1	Which factor is more critical in urban regeneration plans?	12	8	10		30	Frequency
		40	26.66666667	33.33333333		100	Percentage
2	Which approaches are most appropriate for conveying meaning in urban regeneration plans?	4	7	2	17	30	Frequency
		13.33333333	23.33333333	6.666666667	56.66666667	100	Percentage
3	To strengthen the city mindscape, which factor can significantly impact it?	4	7	19		30	Frequency
		13.33333333	23.33333333	63.33333333		100	Percentage
4	Which of the following factors has more priority in the design of bodies and edges in historical cities?	11	9	10		30	Frequency
		36.66666667	30	33.33333333		100	Percentage
5	Which of the following has more priority in urban regeneration projects?	5	1	24		30	Frequency
		16.66666667	3.333333333	80		100	Percentage
6	Which factor can have a more significant impact on the satisfaction of the residents of cities with their place of residence?	7	4	1	18	30	Frequency
		23.33333333	13.33333333	3.333333333	60	100	Percentage
7	To strengthen the social aspects of the landscape, which is preferable in reviving historic cities?	8	3	19		30	Frequency
		26.66666667	10	63.33333333		100	Percentage
8	To strengthen the semantic features of the landscape, which is preferable in reviving historic cities?	23	5	2		30	Frequency
		76.66666667	16.66666667	6.666666667		100	Percentage
9	Which factor can have a more significant effect on promoting a sense of belonging in the residents of cities?	4	5	21		30	Frequency
		13.33333333	16.66666667	70		100	Percentage
10	In your opinion, which factor prioritizes creating a sense of vitality in cities?	10	12	8		30	Frequency
		33.33333333	40	26.66666667		100	Percentage
11	Which factor is more effective in enhancing the perspective of access?	17	8	5		30	Frequency
		56.66666667	26.66666667	16.66666667		100	Percentage
12	To strengthen the city mindscape, which factor can have a more significant impact?	6	5	19		30	Frequency
		20	16.66666667	63.33333333		100	Percentage

Source: Made by the authors

4. To what extent does holding religious and national events in your neighborhood strengthen the sense of belonging to this neighborhood?
    - Very high
    - High
    - Moderate
    - Low
    - Very low
5. How pleasant is it to review the memories of the past in your neighborhood with your neighbors and friends?
    - Very high
    - High
    - Moderate
    - Low
    - Very low
6. To what extent do your shared memories with your friends and neighbors increase your sense of belonging to your neighborhood?
    - Very high
    - High
    - Moderate
    - Low
    - Very low
7. To what extent does having a shared past with neighbors and friends strengthen your willingness to participate in community affairs?
    - Very high
    - High
    - Moderate
    - Low
    - Very low
8. To what extent does the narration of the events and fate of the past and the properties of your place create a sense of pride and belonging in you?
    - Very high
    - High
    - Moderate
    - Low
    - Very low

9. To what extent do you consider a neighborhood organization or board necessary to monitor and control the neighborhood's residents?
    - Very high
    - High
    - Moderate
    - Low
    - Very low
10. If you have capital, how much do you want to invest in your neighborhood?
    - Very high
    - High
    - Moderate
    - Low
    - Very low
11. To what extent do you agree with preserving traditions and customs?
    - Very high
    - High
    - Moderate
    - Low
    - Very low
12. How important is the historical identity of your neighborhood to you?
    - Very high
    - High
    - Moderate
    - Low
    - Very low

Index

A. Moazzeni Khorasgani, *Using Data Science and Landscape Approach to Sustain Historic Cities*, https://doi.org/10.1007/978-3-031-68161-5

MIX
Papier aus verantwortungsvollen Quellen
Paper from responsible sources
FSC® C105338

If you have any concerns about our products,
you can contact us on
ProductSafety@springernature.com

In case Publisher is established outside the EU,
the EU authorized representative is:
Springer Nature Customer Service Center GmbH
Europaplatz 3, 69115 Heidelberg, Germany

Printed by Libri Plureos GmbH
in Hamburg, Germany